W9-CUK-395

OXFORD CHEMISTRY MASTERS

Series Editors

RICHARD G. COMPTON
University of Oxford

STEPHEN G. DAVIES
University of Oxford

JOHN EVANS
University of Southampton

OXFORD CHEMISTRY MASTERS

1 A. Rodger and B. Nordén: *Circular dichroism and linear dichroism*
2 N. K. Terrett: *Combinatorial chemistry*
3 H. M. I. Osborn and T. H. Khan: *Oligosaccharides: Their chemistry and biological roles*
4 H. Cartwright: *Intelligent data analysis in science*
5 S. Mann: *Biomineralization: Principles and concepts in bioinorganic materials chemistry*
6 E. Buncel, R. A. Stairs, and H. Wilson: *The role of the solvent in chemical reactions*

The Role of the Solvent in Chemical Reactions

E. BUNCEL
Queen's University at Kingston, Ontario

R. A. STAIRS
Trent University, Peterborough, Ontario

and

H. WILSON
John Abbott College, Ste. Anne de Bellevue, Quebec

OXFORD
UNIVERSITY PRESS

Chemistry Library

OXFORD
UNIVERSITY PRESS

Great Clarendon Street, Oxford OX2 6DP

Oxford University Press is a department of the University of Oxford.
It furthers the University's objective of excellence in research, scholarship,
and education by publishing worldwide in

Oxford New York

Auckland Bangkok Buenos Aires Cape Town Chennai
Dar es Salaam Delhi Hong Kong Istanbul Karachi Kolkata
Kuala Lumpur Madrid Melbourne Mexico City Mumbai Nairobi
São Paulo Shanghai Taipei Tokyo Toronto

Oxford is a registered trade mark of Oxford University Press
in the UK and in certain other countries

Published in the United States
by Oxford University Press Inc., New York

© Oxford University Press, 2003

The moral rights of the author have been asserted

Database right Oxford University Press (maker)

First published 2003

All rights reserved. No part of this publication may be reproduced,
stored in a retrieval system, or transmitted, in any form or by any means,
without the prior permission in writing of Oxford University Press,
or as expressly permitted by law, or under terms agreed with the appropriate
reprographics rights organization. Enquiries concerning reproduction
outside the scope of the above should be sent to the Rights Department,
Oxford University Press, at the address above

You must not circulate this book in any other binding or cover
and you must impose this same condition on any acquirer

A catalogue record for this title is available from the British Library

Library of Congress Cataloging in Publication Data

Buncel E.
The role of the solvent in chemical reactions / E. Buncel, R.A. Stairs, and H. Wilson.
(Oxford chemistry masters ; 6)
Includes bibliographical references and index.
1. Solvation. 2. Chemical reactions. I. Stairs, R. A. (Robert A.) II. Wilson, H. (Harold)
III. Title IV. Series
QD543 .B945 2003 541.3′4–dc21 2002038142

ISBN 0 19 851100 0 (Pbk)

ISBN 0 19 852760 8 (Hbk)

1 3 5 7 9 10 8 6 4 2

Typeset by Newgen Imaging Systems (P) Ltd., Chennai, India
Printed in Great Britain
on acid-free paper by The Bath Press, Avon

Q D
543
B945
2003
CHEM

Preface

The role of the solvent in chemical reactions is one of immediate and daily concern to the practising chemist. Whether in the laboratory or in industry, most reactions are carried out in the liquid phase. In the majority of these, one or two reacting components, or reagents, with or without a catalyst, are dissolved in a suitable medium and the reaction is allowed to take place. The exceptions, some of which are of great industrial importance, are those reactions taking place entirely in the gas phase or at gas–solid interfaces, or entirely in solid phases. Reactions in the absence of solvent are rare, though they include such important examples as bulk polymerization of styrene or methyl methacrylate. Of course, one could argue that the reactants are their own solvent.

Given the importance of solvent, the need for an in-depth understanding of a number of cognate aspects seems obvious. In the past, many texts of inorganic and organic chemistry either did not bother to mention that a given reaction takes place in a particular solvent, or they mentioned the solvent only in a perfunctory way. Explicit discussion of the effect of changing the solvent was rare, but this is changing. Recent texts, for example, Streitwieser *et al.* (1992), Carey (1996), Solomons and Fryhle (2000), and Claydon *et al.* (2001), devote at least a few pages to solvent effects. Morrison and Boyd (1992) and Huheey *et al.* (1993) each devote a whole chapter to the topic.

It is the aim of this monograph to amplify these brief treatments, and so to bring the role of the solvent to the fore at an early stage of the student's career. Chapter 1 begins with a general introduction to solvents and their uses. While it is assumed that the student has taken courses in the essentials of thermodynamics and kinetics, we make no apology for continuing with a brief review of essential aspects of these concepts. The approach throughout is semiquantitative, neither quite elementary nor fully advanced. We have not avoided necessary mathematics, but have made no attempt at rigour, preferring to outline the development of unfamiliar formulas only in sufficient detail to avoid mystification.

The physical properties of solvents are first considered in Chapter 2, entitled 'The Solvent as Medium', which highlights, for example, Hildebrand's solubility parameter, and the Born and Kirkwood–Onsager electrostatic theories. An introduction to empirical parameters is also included. Chapter 3, 'The Solvent as Participant', deals chiefly with the ideas of acidity and basicity and the different forms in which they may be expressed. Given the complexities surrounding the subject, the student is introduced in Chapter 4 to empirical correlations of solvent properties. In the absence of complete understanding of solvent behaviour, one comes to appreciate the attempts that have been made by statistical analysis (*chemometrics*) to rationalize the subject. A more theoretical approach is made in Chapter 5, but even though this is entitled 'Theoretical Calculations', there is in fact no rigorous theory presented. Nevertheless, the interested student may be sufficiently motivated to follow up on this topic. Chapters 6 and 7 deal with some specific examples of solvents: dipolar-aprotic solvents like dimethylformamide and dimethyl sulfoxide and more common acidic/basic solvents, as well as chiral solvents and the currently highlighted room-temperature ionic liquids. The monograph ends with an appendix, containing general tables. These include a table of physical properties of assorted solvents, with some notes on safe handling and disposal of wastes, lists of derived and empirical parameters, and a limited list of their values.

A few problems have been provided for some of the chapters.

We were fortunate in being able to consult a number of colleagues and students, including (in alphabetical order) Peter F. Barrett, Natalie M. Cann, Doreen Churchill, Robin A. Cox, Julian Dust, Robin Ellis, Errol G. Lewars, Lakshmi Murthy, Igor Svishchev, Christian Reichardt, and Matthew Thompson, who have variously commented on early drafts of the text, helped us find suitable examples and references, helped with computer problems, and corrected some of our worst errors. They all have our thanks.

Lastly, in expressing our acknowledgements we wish to give credit and our thanks to Professor Christian Reichardt, who has written the definitive text in this area with the title *Solvents and Solvent Effects in Organic Chemistry* (2nd edn. 1988, 534 pages; 3rd edn. 2003, 629 pages). It has been an inspiration to us to read this text and on many occasions we have been guided by its authoritative and comprehensive treatment. It is our hope that having read our much shorter and more elementary monograph, the student will go to Reichardt's text for deeper insight.

E. B., Kingston, *Ontario*
R. A. S., Peterborough, *Ontario*
H. W., Montreal, *Quebec*

February 2003

Contents

1 Introduction

1.1 Generalities

The alchemists' adage *Corpora non agunt nisi fluida*, 'Substances do not react unless fluid', while not strictly accurate (for crystals can be transformed by processes of nucleation and growth), is still generally true enough to be worthy of attention. Seltzer tablets, for instance, must be dissolved in water before they will react to evolve carbon dioxide. The 'fluid' state may be gaseous or liquid, and the reaction may be a homogeneous one, occurring throughout a single gas or liquid phase, or a heterogeneous one, occurring only at an interface between a solid and a fluid, or between two immiscible fluids. As the title suggests, this book is concerned mainly with homogeneous reactions, and will emphasize reactions of substances dissolved in liquids of various kinds.

The word 'solvent' implies that the component of the solution so described is present in excess; one definition is 'the component of a solution that is present in the largest amount'. In most of what follows it will be assumed that the solution is 'dilute'. We will not attempt to define how dilute is 'dilute', except to note that we will routinely use most physicochemical laws in their simplest available forms, and then require that all solute concentrations be low enough that the laws are valid, at least approximately. See, for example, Problem 2.1.

Of all solvents, water is of course the cheapest and closest to hand. Because of this alone it will be the solvent of choice for many applications. In fact, it has dominated our thinking for so long that any other solvent tends to be tagged *non-aqueous*, as if water were in some essential way unique. It is true that it has an unusual combination of properties, but so have many others. More and more, however, other solvents are coming into use in the laboratory and in industry. Aside from organic solvents such as methanol, acetone, and hydrocarbons, which have been in use for many years, industrial processes use such solvents as sulphuric acid, hydrogen fluoride, ammonia, molten sodium hexafluoroaluminate (cryolite), various other 'ionic liquids' (Welton 1999), and liquid metals. Jander and Lafrenz (1970) cite the industrial use of bromine to separate caesium bromide (solubility 19.3 g $100\,g^{-1}$ bromine) from the much less soluble rubidium salt. The list of solvents available for preparative and analytical purposes in the laboratory now is long and growing, and though water will still be the first solvent that comes to mind, there is no reason to stop there.

The study of solvent effects was at first largely the work of physical-organic chemists, notably Hughes and Ingold (1935) and Grunwald and Winstein (1948). One of us (R. A. S.) was privileged to attend Ingold's lectures at Cornell that became the basis of his book (Ingold 1969), while E. B. can still recall vividly the undergraduate lectures by both Hughes and Ingold on the effect of solvent in nucleophilic substitution: the

Hughes–Ingold Rules (Ingold 1969). Inorganic chemists soon followed. Tobe and Burgess (1999: 335) remark that while inorganic substitution reactions of known mechanism were used to probe solvation and the effects of solvent structure, medium effects have been important in understanding the mechanisms of electron transfer.

If a solvent is to be chosen for the purpose of preparation of a pure substance by synthesis, clearly the solvent must be one that will not destroy the desired product, or transform it in any undesirable way. Usually it is obvious what must be avoided. For instance, one would not expect to be able to prepare a strictly anhydrous salt using water as the reaction medium. Anhydrous chromium(III) chloride must be prepared by some reaction that involves no water at all, neither in a solvent mixture nor in any of the starting materials, nor as a byproduct of reaction. A method that works uses the reaction at high temperature of chromium(III) oxide with tetrachloromethane (carbon tetrachloride), according to the equation:

$$Cr_2O_3(s) + 3CCl_4(g) \rightarrow 2CrCl_3(s) + 3COCl_2(g)$$

Here no solvent is used at all.[1] Some other anhydrous salts may be prepared using such solvents as sulphur dioxide, dry diethyl ether (a familiar example is the Grignard reaction, in which mixed halide-organic salts of magnesium are prepared as intermediates in organic syntheses), hydrogen fluoride, etc.

A more subtle problem is to maximize the yield of a reaction that could be carried out in any of a number of media. Should a reaction be done in a solvent in which the desired product is most or least soluble, for instance? The answer is not immediately clear. In fact one must say, 'It depends . . . .' If the reaction is between ions of two soluble salts, the product will precipitate out of solution if it is insoluble. For example, a reaction mixture containing barium, silver, chloride, and nitrate ions will precipitate insoluble silver chloride if the solvent is water, but in liquid ammonia the precipitate is barium chloride. Another example, from organic chemistry, described by Collard *et al.* (2001) as an experiment suitable for an undergraduate laboratory, is the dehydrative condensation of benzaldehyde with pentaerythritol in aqueous acid to yield the cyclic acetal, 5,5-bis(hydroxymethyl-2-phenyl-1,3-dioxane), **1**:

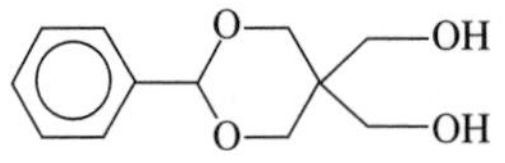

O OH
O OH

1

At 30 °C the product is sufficiently insoluble to appear as a precipitate, so the reaction proceeds in spite of the formation of water as byproduct.

On the other hand, we will show in Chapter 2 that, in a situation where all the substances involved in a reaction among molecules are more or less soluble, *the most soluble substances will be favoured at equilibrium.*

[1] Caution: The reagent tetrachloromethane and the byproduct phosgene are toxic and environmentally undesirable.

1.2 Classification of solvents

Solvents may be classified according to their physical and chemical properties at several levels. The most striking differences among liquids that could be used as solvents are observed between *molecular liquids*, *ionic liquids* (molten salts or salt mixtures) and *metals*. They can be considered as extreme types, and represented as the three vertices of a triangle (Trémillon 1974) (see Fig. 1.1). Intermediate types can then be located along edges or within the triangle. Concentrated aqueous salt solutions lie along the molecular-ionic edge, for instance.

Among the molecular liquids, further division based on physical and chemical properties leads to categories variously described (Barthel and Gores 1994; Reichardt 1988) as *inert* (unreactive, with low or zero dipole moments and low polarizability), *inert-polarizable* (e.g. aromatics, polyhalogenated hydrocarbons), *protogenic* (hydrogen-bonding proton donors, HBD), *protophilic* (hydrogen-bonding proton acceptors, HBA), *amphiprotic* (having both HBD and HBA capabilities) and *dipolar-aprotic* (having no marked HBD or HBA tendencies, but possessing substantial dipole moments). Examples of these classes are listed in Table 1.1. The ability of solvent molecules to act as donors or acceptors of electron pairs, that is, as Lewis bases or acids, complicates the classification. Nitriles, ethers, dialkyl sulphides, and ketones are electron-pair donors (EPD), for example; sulphur dioxide and tetracyanoethene are electron-pair acceptors (EPA). EPD and EPA solvents can be further classified as *soft* or *hard* (classifying can be habit-forming). Pushing the conditions can cause normally inert substances to show weak prototropic properties: dimethyl sulphoxide can lose a proton to form the *dimsyl* anion, $CH_3SOCH_2^-$ in very strongly basic media (Olah *et al.* 1985). An equilibrium concentration of dimsyl, very small, though sufficient for hydrogen–deuterium isotopic exchange to occur between dimethyl sulphoxide and D_2O, is set up even in very dilute aqueous NaOH (Buncel *et al.* 1965). Carbon monoxide, not normally considered a Brønsted base, can

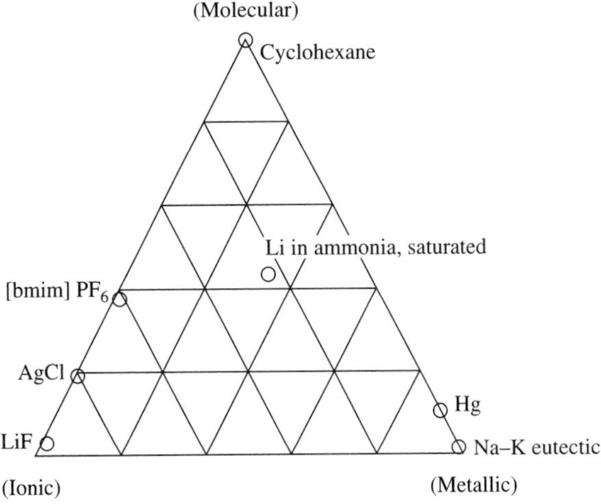

Fig. 1.1 Ternary diagram for classification of liquids—schematic—location of points is conjectural (After Trémillon (1974) [bmim]PF_6 represents a room-temperature ionic liquid (see Chapter 7).

Table 1.1 Molecular solvents

Classes	Examples
Inert	Aliphatic hydrocarbons, fluorocarbons
Inert-polarizable	Benzene (π-EPD), tetrachloromethane, carbon disulphide, tetracyanoethene (π-EPA)
Protogenic (HBD)	Trichloromethane
Protophilic (HBA)	Tertiary amines (EPD)
Amphiprotic	Water, alcohols; ammonia is more protophilic than protogenic, while acetic acid is the reverse
Dipolar-aprotic	Dimethylformamide, acetonitrile (EPD, weak HBA), dimethyl sulphoxide, hexamethylphosphortriamide

be protonated in the very strongly acidic medium, HF–SbF_5 (de Rege *et al.* 1997).

1.3 Some essential thermodynamics and kinetics: tendency and rate

How a particular reaction goes or does not go in given circumstances depends on two factors, which may be likened, 'psychochemically' speaking, to 'wishing' and 'being able'.[2] The first is the *tendency to proceed*, or the degree to which the reaction is out of equilibrium, and is related to the equilibrium constant and to free energy changes (Gibbs or Helmholtz). It is the subject of *chemical thermodynamics*. The second is the speed or *rate* at which the reaction goes, and is discussed in terms of rate laws, mechanisms, activation energies, etc. It is the subject of *chemical kinetics*. We will need to examine reactions from both points of view, so the remainder of this chapter will be devoted to reviewing the essentials of these two disciplines, as far as they are relevant to our needs. The reader may wish to consult, for example, Atkins (1998) for fuller discussions of relevant thermodynamics and kinetics.

1.4 Equilibrium considerations

For a system at constant pressure, which is the usual situation in the laboratory when we are working with solutions in open beakers or flasks, the simplest formulas to describe equilibrium are written in terms of the Gibbs

[2] There is a word, very pleasing to us procrastinators, 'velleity', which is defined (Fowler *et al.* 1976) as 'low degree of volition not prompting to action.' See also Ogden Nash (1938).

energy G, and the enthalpy H. For a reaction having an equilibrium constant K at the temperature T, one may write:

$$\Delta G_T^0 = -RT \ln K \tag{1.1}$$

$$\Delta H_T^0 = -R\left(\frac{\partial \ln K}{\partial(1/T)}\right)_P \tag{1.2}$$

The equilibrium constant K is of course a function of the activities of the reactants and products, for example, for a reaction

$$\mathrm{A} + \mathrm{B} \rightleftharpoons \mathrm{Y}$$

$$K = \frac{a_\mathrm{Y}}{a_\mathrm{A} \cdot a_\mathrm{B}} \tag{1.3}$$

By choice of standard states one may express the activities on different scales. For reactions in the gas phase, it is convenient, and therefore common, to choose a standard state of unit activity on a scale of pressure such that the limit of the value of the dimensionless activity coefficient, $\gamma = a_i/P_i$, as the pressure becomes very low, is unity. The activity on this scale is expressed in pressure units, usually atmospheres or bars, so we may write

$$K = \frac{\gamma_\mathrm{Y} P_\mathrm{Y}}{\gamma_\mathrm{A} P_\mathrm{A} \gamma_\mathrm{B} P_\mathrm{B}} = \Gamma_\gamma K_P \tag{1.4}$$

The activity-coefficient quotient Γ_γ is unity for systems involving only ideal gases, and for real gases at low pressure.

For reactions involving only condensed phases, including those occurring in liquid solutions, which are our chief concern, the situation is very different. Three choices of standard state are in common use. For the solvent (i.e. the substance present in largest amount), the standard state almost universally chosen is the pure liquid. This choice is also often made for other liquid substances that are totally or largely miscible with the solvent. The activity scale is then related to the mole fraction, through the *rational activity coefficient* f, which is unity for each pure substance. For other solutes, especially those that are solid when pure, or for ionic species in solution in a non-ionic liquid, activity scales are used that are related either to the molar concentration or the molality, depending on experimental convenience. On these scales, the activity coefficients become unity in the limit of low concentration.

If a substance present in solution is to some extent volatile, that is, if it exerts a measurable vapour pressure, its activity in solution can be related to its activity in the gas (vapour) phase. If the solution is ideal, all components obey Raoult's Law, expressed by eqn 1.5, and illustrated by the dashed lines in Fig. 1.2.

$$p_i = p_i^0 x_i \tag{1.5}$$

Here p_i is the vapour pressure of the ith substance over the solution, p_i^0 is the vapour pressure it would exert in its standard (pure liquid) state, and x_i is its mole fraction in the solution. We can now define an 'absolute' activity (not

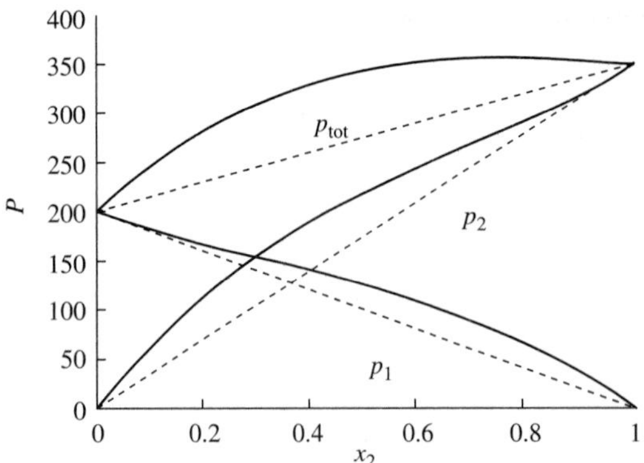

Fig. 1.2 Vapour pressure over binary solutions. Dashed lines: ideal (Raoult's Law); Solid curves: positive deviations from Raoult's Law. Note that where $x_2 \ll 1$, P_1 is close to ideal, and vice versa.

really absolute, but relative to the gas phase standard state on the pressure scale as above) measured by p_i, assuming that the vapour may be treated as an ideal gas or by the *fugacity*[3] if necessary. We will always make the '*ideal gas*' *assumption*, without restating it.

1.5 Thermodynamic transfer functions

The thermodynamic equilibrium constant as defined above is independent of the solvent. The practical equilibrium constant is not, because the activity coefficients of the various reactant and product species will change in different ways when the reaction is transferred from one solvent to another. One way of considering these changes is through the use of thermodynamic transfer functions. The standard Gibbs energy of a reaction in a solvent **S**, ΔG^0_S, may be related to that in a reference solvent **0**, ΔG^0_0, by considering the change in Gibbs energy on transferring each reactant and product species from the reference solvent to **S**. The reference solvent may be water, or the gas phase (no solvent). Other functions (enthalpy, entropy) can be substituted for G. A reaction converting reactants **R** to products **P** in the two solvents can be represented in a Born–Haber cycle:

$$\begin{array}{llcl}
(\textit{In}\ \mathbf{S}) & \mathrm{R(S)} & \overset{\Delta G^0_\mathrm{S}}{\rightleftharpoons} & \mathrm{P(S)} \\
\delta_\mathrm{tr}G^{(\mathrm{R})} & \Uparrow & & \Uparrow\ \delta_\mathrm{tr}G^{(\mathrm{P})} \\
(\textit{In}\ \mathbf{0}) & \mathrm{R(0)} & \overset{\Delta G^0_0}{\rightleftharpoons} & \mathrm{P(0)}
\end{array}$$

$$\Delta G^0_\mathrm{S} = -\delta_\mathrm{tr}G^{(\mathrm{R})} + \Delta G^0_0 + \delta_\mathrm{tr}G^{(\mathrm{P})} \tag{1.6}$$

For each participating substance **i**, the transfer function term $\delta_\mathrm{tr}G^{(\mathrm{i})}$ can be obtained from vapour pressure, solubility, electrical potential, or other

[3] Fugacity f is pressure corrected for non-ideality. It is defined so that the Gibbs energy change on isothermal, reversible expansion of a mole of a real gas, $\Delta G = \int V\,\mathrm{d}P = RT\ln(f/f_0)$. For a real gas at low enough pressures, $f = P$. Fugacities can be calculated from the equation of state of the gas if needed. See any physical chemistry textbook, for example, Atkins (1998: 133–136). For an only slightly non-ideal gas $f = P^2V_m/RT$, approximately.

measurements that enable the calculation of activity coefficients and hence of standard Gibbs energies, using eqn 1.7.

$$\delta G_{\mathrm{tr}}^{(i)} = G_{\mathrm{S}}^{0(i)} - G_{0}^{0(i)} \quad (1.7)$$

Since the Gibbs energy and the activity coefficient are related through eqn 1.8, the above development could have been carried out in terms of ln a_i or ln f_i.

$$G_{m,i} - G_{m,i}^{0} = RT \ln(a/a_i^0) = RT \ln(f/f_i^0) \quad (1.8)$$

Because of the analogy between the transition state in kinetics and the products in equilibrium (see Section 1.6), similar considerations can be applied to the understanding of solvent effects on reaction rates. This will be illustrated in Chapter 6.

1.6 Kinetic considerations: Collision theory

Elementary reactions occurring in the gas phase have been fruitfully discussed in terms derived from the Kinetic-Molecular Theory of Gases. The result is eqn 1.9,

$$\text{rate} = PZ_0[\mathrm{A}][\mathrm{B}]e^{-E_{\mathrm{a}}/RT} \quad (1.9)$$

$$Z_0 = \pi d^2 \left(\frac{8kT}{\pi\mu}\right)^{1/2} N^2 \quad (1.10)$$

where Z_0 is the number of collisions per unit time between A and B molecules at unit concentrations, [A] and [B] represent the concentrations of the reacting species, d is the mean diameter of A and B, μ their reduced mass, and E_{a} is the activation energy. P is the steric or probability factor, that is, the probability that the colliding molecules are in suitable orientations and internal configuration to permit reaction, as illustrated in Fig. 1.3. The factors PZ_0 are usually combined to form the *Arrhenius pre-exponential factor*, usually denoted by A. Equations 1.9 and 1.10 have allowed a substantial level of understanding of simple reactions to be achieved, and by combining elementary steps into multistep mechanisms, complex reactions may also be described. This simple Arrhenius treatment is not applicable to reactions in solution, however, so for our purposes another approach is needed.

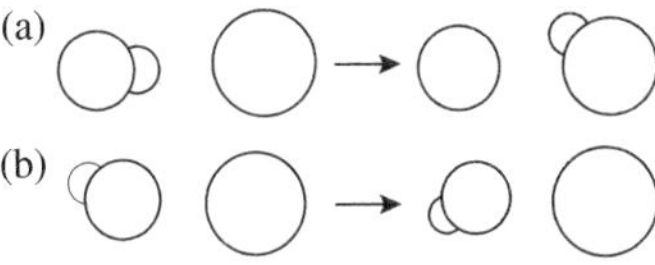

Fig. 1.3 Successful (a) and unsuccessful (b) transfer of a hydrogen atom.

1.7 Transition-state theory

The variously-named *transition-state theory* (the preferred name) or *absolute reaction rate theory*, developed by Eyring and associates (Eyring 1935; Laidler and Meiser 1995: 382–7; Berry *et al.* 2000: 911–27) and by Evans and Polanyi (1935), takes a quite different view. The reacting molecules are considered as forming an 'activated complex', or 'transition state', that resembles an ordinary molecule in all respects but one, which is that one of its normal modes of vibration is *not* a vibration, because there is no restoring force; rather it will lead to decomposition of the complex, either to form the products of the reaction or to reform the starting molecules. Quantum-mechanical calculations of the energetics and geometry of molecules in configurations that represent transition states can be carried out using such computer programs as GAUSSIAN, SPARTAN, or HYPERCHEM (Levine 2000). Of the normal modes of vibration of such a transition-state 'molecule', one has a *negative force constant*. What is meant by this is that there is no force restoring the molecule to an equilibrium configuration in the direction of this motion; in fact the force is repulsive, leading to rearrangement or decomposition, to form the products of the reaction, or to reform the starting molecules. Since the force constant is negative, the frequency, which depends on the square root of the force constant, contains the factor $\sqrt{-1}$, that is, it is imaginary. A graph of the energy of the system as a function of the normal coordinates of the atoms (the *potential energy surface*) in the vicinity of the transition state takes the form of a *saddle* or *col*, illustrated in Fig. 1.4.

From the saddle point, the energy increases in all the principal directions except along the direction that leads to reaction (forward) or (backward) to reform the starting materials. The course of a simple reaction may be represented as motion along the *reaction coordinate*, which is a combination of atomic coordinates leading from the initial configuration (reactants) through the transition state to the final configuration (products) along, or nearly along, the path of least energy. Figure 1.5(a) shows a projection of the path of least energy on the potential energy surface for a very simple reaction, in which a hydrogen atom attacks a hydrogen molecule directly at one end, and one atom is transferred to the attacking atom. The reaction coordinate is measured along the (approximately hyperbolic) pathway. The energy as a function of the reaction coordinate is shown in Fig. 1.5(b).

In most reactions, especially those taking place in solution, the situation is more complicated. For instance, Fig. 1.6 shows a possible form of the energy profile for a reaction in which one ligand in a transition-metal complex ion is

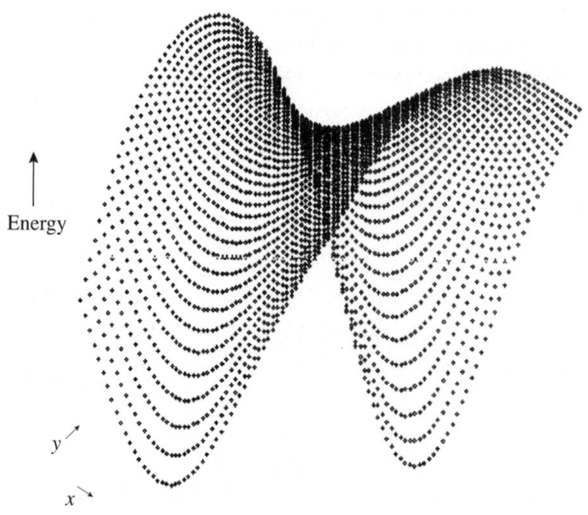

Fig. 1.4 A portion of a potential-energy surface $E(x,y)$, showing a saddle point.

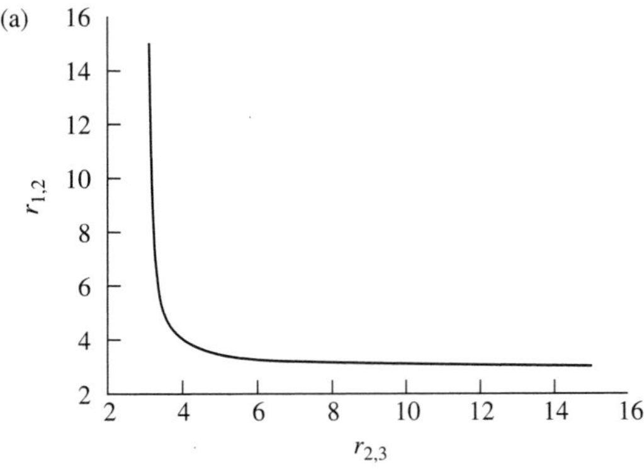

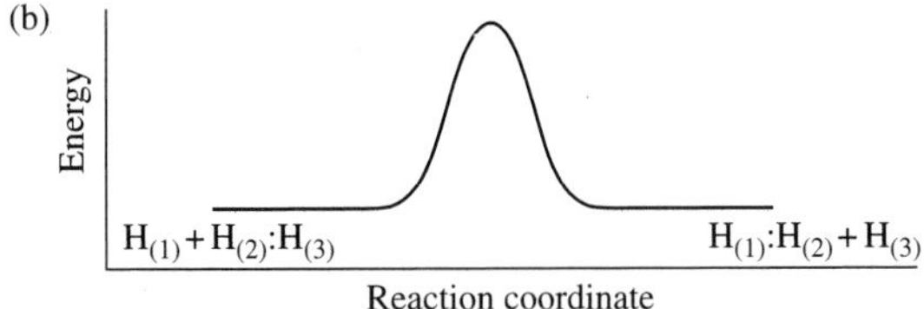

Fig. 1.5 (a) The reaction pathway of least energy and (b) the profile along the pathway for the hydrogen atom–molecule exchange reaction (schematic).

replaced by another. In the scheme below, M represents a trivalent metal ion. There may first be formed an outer-sphere complex, perhaps an ion pair (step 1.11), which then rearranges (step 1.12) so the arriving and leaving ligands change places (not necessarily with retention of configuration). The leaving ligand, now in the outer sphere, finally leaves (step 1.13).

$$X^- + M(H_2O)_5Y^{2+} = X^-M^{(III)}(H_2O)_5Y^{2+} \tag{1.11}$$

$$X^-M(H_2O)_5Y^{2+} \rightarrow [M^{(III)}(H_2O)_5XY]^+ \rightarrow Y^-M(H_2O)_5X^{2+} \tag{1.12}$$

(The seven-coordinate species may be a true transition state, corresponding to the curve with a single maximum in Fig. 1.6, or a transient intermediate, corresponding to the light curve with a dip between two maxima.)

$$Y^-M(H_2O)_5X^{2-} \rightarrow M(H_2O)_5X^{2-} + Y^- \tag{1.13}$$

The composition of the activated complex may be deduced from the rate law. In the case of a multi-step reaction, if one step is rate controlling, which is usually true, the composition of the activated complex of the rate-controlling step may still be deduced from the rate law for the overall reaction. For example, if a reaction between two substances A and B follows a rate law of the form of eqn 1.14 (over certain ranges of concentrations and temperatures):

$$\text{rate} = k[A]^n[B]^m \tag{1.14}$$

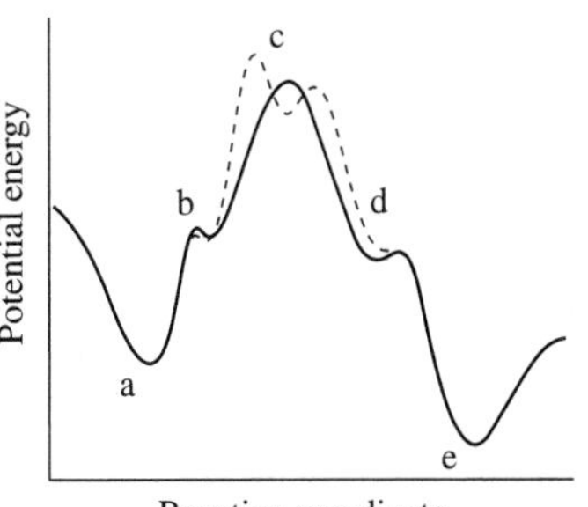

Fig. 1.6 A possible, more realistic reaction profile for a ligand-exchange reaction, showing reactants (a), precursor (b) and successor (d) complexes, the possibility of the formation of a reactive intermediate (c), and products (e). Redrawn after Kettle (1996), with permission.

the activated complex has a composition represented by A_nB_m (there are some subtle aspects of this rule; e.g. see, Problem 2.2). We know nothing, however, of its structure, nor of the steps in the reaction, in the absence of other evidence. Nevertheless, one may write a statement resembling an equilibrium-constant expression, relating the activities or, approximately, the concentrations of the reactants and the activated complex (represented by the 'double-dagger' or 'Cross of Lorraine' symbol ‡):

$$K^{\ddagger} = \frac{[\ddagger]}{[\mathrm{A}]^n[\mathrm{B}]^m} \tag{1.15}$$

Then, if we assume that the rate of decomposition of the complex is first order, that is, that it reacts to form the products at a rate proportional to its concentration, we obtain:

$$\text{rate} \propto [\ddagger] \propto K^{\ddagger}[\mathrm{A}]^n[\mathrm{B}]^m \tag{1.16}$$

that is, the ordinary rate constant, k, is proportional to $K^{\ddagger}$.

Specifically, one may write (Laidler and Meiser 1995: 741; Atkins 1998: 832):

$$k = \kappa\frac{k_0 T}{h}K^{\ddagger} \tag{1.17}$$

where k_0 is Boltzmann's constant, h is Planck's constant and T is the absolute temperature. The group $\kappa k_0T/h$ has a value of about $6\,\mathrm{ps}^{-1}$ at 298 K. The factor κ is a constant, the *transmission coefficient*, the value of which is close to unity for bimolecular reactions in the gas phase. Abboud *et al.* (1993: 75) cite a computer simulation study (Wilson 1989) of the chloride exchange reaction:

$${}^{*}\mathrm{Cl}^- + \mathrm{CH_3\text{–}Cl} \rightarrow {}^{*}\mathrm{Cl\text{–}CH_3} + \mathrm{Cl}^-$$

in which the transmission coefficient was calculated to be unity for the reaction in the gas phase, but 0.55 in aqueous solution, apparently owing to confinement of the reacting species within a 'cage' of water molecules, so that multiple crossings of the transition barrier can occur.

To the extent that $K^{\ddagger}$ can be considered an ordinary equilibrium constant, one may then apply the usual thermodynamic relations, that is

$$\Delta_{\ddagger}G^0 = -RT\ln K^{\ddagger} \tag{1.18}$$

$$\Delta_{\ddagger}H^0 = -R\left[\frac{\partial \ln K^{\ddagger}}{\partial(1/T)}\right]_p \tag{1.19}$$

$$\Delta_{\ddagger}S^0 = \frac{\Delta_{\ddagger}H^0 - \Delta_{\ddagger}G^0}{T} \tag{1.20}$$

Equation 1.17 may then be written:

$$k = \kappa \frac{k_0 T}{h} e^{-\Delta_{\ddagger} G^0/RT} = \kappa \frac{k_0 T}{h} e^{\Delta_{\ddagger} S^0/R} e^{-\Delta_{\ddagger} H^0/RT} \tag{1.21}$$

or, in logarithmic form,

$$\ln k = \ln(\kappa k_0/h) + \frac{\Delta_{\ddagger} S^0}{R} + \ln T - \frac{\Delta_{\ddagger} H^0}{RT} \tag{1.22}$$

and differentiating with respect to $(1/T)$,

$$\frac{\mathrm{d}(\ln k)}{\mathrm{d}(1/T)} = -T - \frac{\Delta_{\ddagger} H^0}{T} = -\frac{\Delta_{\ddagger} H^0 + RT}{R} = -\frac{E_\mathrm{a}}{R} \tag{1.23}$$

where E_a is the ordinary (Arrhenius) experimental activation energy, which is thus equal to $\Delta_{\ddagger} H^0 + RT$.

The two theoretical approaches, one in terms of molecular collisions and the other in terms of an activated complex, are not opposed, but complementary. A key to the connection between them is the entropy of activation. When both the rate constant and the temperature coefficient of the rate constant are known, $\Delta_{\ddagger} G^0$ and $\Delta_{\ddagger} H^0$ $(=E_a - RT)$ can be used with eqn 1.20 to obtain $\Delta_{\ddagger} S^0$. In an ordinary bimolecular reaction with no special steric requirements, the formation of the activated complex means the formation of one rather 'loose' molecule from two. A negative entropy change is to be expected, perhaps comparable to that for the combination of two iodine atoms (Atkins 1998: 926),

$$2\mathrm{I(g)} \rightleftharpoons \mathrm{I_2(g)}: \quad \Delta S^0_{298} = -100.9\,\mathrm{J^{-1}\,mol^{-1}}$$

A value much more negative than this implies the loss of much freedom of motion on formation of the complex, and corresponds to a small value of P, the steric factor in the collision theory. On the other hand, less negative or even positive values of $\Delta_{\ddagger} S^0$ occasionally occur, though rarely if ever for bimolecular reactions in the gas phase. They imply that the complex is very loosely bound, or, in solution, that the complex is less tightly solvated than are the reactant species.

Most reactions in the gas phase at low pressures can be treated as if no foreign molecules (i.e. other than reactants, intermediates, or products of the reaction) are present. Thus the presence of an inert gas such as argon is not important. An exception to this rule is any reaction in which two atoms combine to form a stable diatomic molecule. This cannot happen unless some means exists of getting rid of the energy of formation of the bond. A *third body*, which may be any molecule or the container wall, must be present to absorb some of this energy. Its function has been likened (more 'psychochemistry'?) to that of a chaperon (Laidler 1987: 183), present, not to prevent union, but to ensure that the union is stable and is not formed in an excited state. An extensive literature exists on the efficacy of different molecules as third bodies (Troe 1978; Mitchell 1992), and on the influence of the container wall in this and other ways.

1.8 Reactions in solution

When it comes to reactions in solution, the results of kinetic experiments are difficult to understand, except qualitatively, through the collision theory. The very concept of a collision is hard to define in the liquid phase, in which molecules are not free to travel in straight lines between collisions, but move in constant contact with neighbours, in a 'tipsy reel' (J. H. Hildebrand's phrase). What happens when two solute molecules come into contact in solution, perhaps to react, or perhaps to diffuse apart unchanged, is sometimes called an 'encounter', rather than a collision. Rabinowitch and Wood (1936) demonstrated this by the use of a model in which a few metal balls rolled about on a level table, making collisions which were detected electrically. When many non-conducting balls were added to the set on the table, so that it became rather crowded, instead of single collisions at long and irregular intervals, collisions happened in groups, while the two metal balls were temporarily trapped in a cage of other balls. Computer modelling in three dimensions, using simulated hard spheres, gave a similar result: collisions in a crowded space between labelled molecules occurred in groups of from ten to nearly a hundred, depending on the degree of crowding. In the hard sphere representation, collisions could still be recognized. In a more realistic computer model in which molecular attractions and repulsions are both dependent on distance (which enormously increases the amount of calculation required), an encounter would become a continuous interaction of a complicated kind. During the encounter, something resembling a definite complex, called an *encounter complex*, is present (Langford and Tong 1977). Eigen and Tamm (1962), in work on ultrasonic effects on solutions of sulfates of divalent metals, interpreted their data as showing that such an encounter complex was formed between the oppositely-charged ions, but an encounter complex may exist in the absence of such electrostatic assistance.

1.9 Diffusion-controlled reactions

Consider a bimolecular reaction in solution as occurring in two steps. In the first step, an encounter complex is formed:

$$\mathrm{A} + \mathrm{B} \rightarrow [\mathrm{AB}] \quad k_1$$

The complex may then either revert to separated reactants, or react to form products:

$$[\mathrm{AB}] \rightarrow \mathrm{A} + \mathrm{B} \quad k_{-1}$$

$$[\mathrm{AB}] \rightarrow \mathrm{P} \quad k_2$$

Applying the steady-state assumption to the concentration of the encounter complex:

$$\frac{d[AB]}{dt} = k_1[A][B] - k_{-1}[AB] - k_2[AB] = 0 \tag{1.24}$$

it may be shown that the rate of formation of products is given by eqn 1.25:

$$\frac{d[P]}{dt} = k_2[AB] = k[A][B]; \quad k = \frac{k_2 k_1}{k_2 + k_{-1}} \tag{1.25}$$

If the encounter complex reacts to form products much faster than it reverts to reactants, that is, if $k_2 \gg k_{-1}$, then $k \simeq k_1 k_2 / k_2 = k_1$, that is, the rate is controlled by the rate of formation of the encounter complex. Such a reaction is described as *diffusion-controlled* or *encounter-controlled*. The magnitude of k_1 is approximately given by eqn 1.26 (Cox 1994: 59; Atkins 1998: 826):

$$k_1 \simeq \frac{8000RT}{3\eta} \tag{1.26}$$

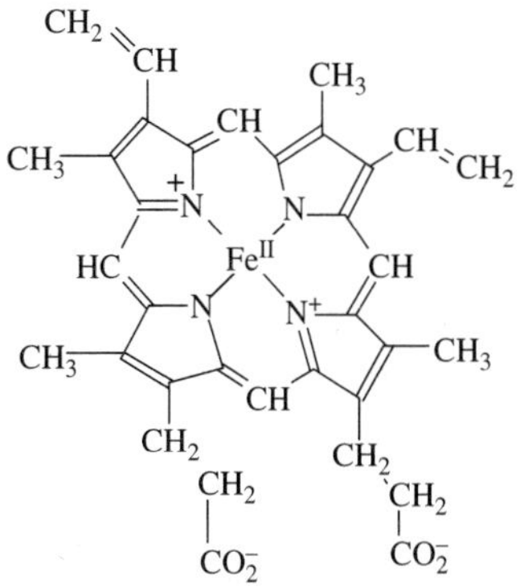

where R is the gas constant and η is the viscosity. A factor of 1000 lets the result be in the conventional units, $\mathrm{L\,mol^{-1}\,s^{-1}}$. A reaction, the rate of which is dependent on bond making or breaking when run in an ordinary solvent, may be diffusion controlled when run in such a highly viscous solvent as glycerol (1,2,3-propanetriol). This has been demonstrated with reactions as diverse as solvent exchange in complexes of Cr^{2+}, Cu^{2+}, and Ni^{2+} (Caldin and Grant 1973) and reactions of ferroprotoporphyrin IX **2** with CO and with O_2 (Caldin and Hasinoff 1978).

1.10 Reaction in solution and the Transition-State Theory

The most satisfactory way to consider reactions in solution is through the thermodynamic interpretation of the Transition-State Theory, by examining the effects of various properties of the solvent on the activity of each reactant species and on the activated complex, treating the latter almost as 'just another molecule'. The solvent can influence the solute molecules by acting on them with 'physical' forces (van der Waals forces and electrostatic forces due to the polarity and polarizability of solvent and solute molecules), but also in more obviously 'chemical' ways, through the formation of hydrogen bonds or molecular or ionic complexes of various kinds. Changing from an 'inert' solvent, one that solvates solutes weakly, to one that exerts stronger forces, may either retard or accelerate a reaction through the change in enthalpy of activation. This depends on whether the latter solvent interacts more strongly with the reactants or with the activated complex. Dewar (1992) discusses an example of a reaction in which the necessity of desolvation of an attacking ion has a profound effect (incidentally, he describes (p. 160) the hard–soft acid–base distinction as 'mythical', at least as an explanation of the difference between nucleophilic substitutions at carbonyl and saturated carbon atoms). To take the simplest case, if the reactant in a unimolecular reaction (perhaps an isomerization or an S_N1 substitution) is more strongly solvated, the reaction will be retarded, through the increase in activation enthalpy; if it is the activated complex that is more solvated, the reverse effect will be found (see Fig. 1.7).

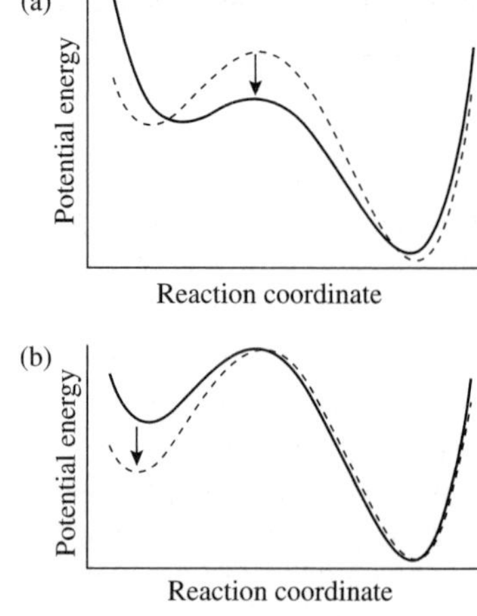

Fig. 1.7 Effect of complexation on activation energy: Dashed curves represent the energy profile in the absence of complexation. (a) Complexation of the activated complex (reduced activation energy). (b) Complexation of the reactant (increased activation energy).

Energy diagrams such as Figure 1.7, represent a cross-cut along the reaction coordinate of the full potential energy surface, i.e. the lowest energy pathway. The 'reaction coordinate' axis in these diagrams represents the various changes in interatomic distance implicit in bond formation and bond rupture, as well as any accompanying changes in bond angles. Different 2-D simplifications of the 3-D potential energy surface have been proposed, for example, representing the energy-coordinate axis as either with or without contour lines. A popular model of such a potential energy surface diagram is the so-called More O'Ferrall-Jencks, four-cornered diagram (More O'Ferrall 1970; Jencks 1977), widely used to illustrate S_N1, S_N2 and borderline mechanisms, as well as reactions at carbonyl centres illustrating general acid-general base catalysis mechanisms (Jencks 1987; Buncel *et al.* 1980, 1982; Richard 1995). An illustration of the More O'Ferrall-Jencks energy diagram representing the S_N1/S_N2 processes, including contours for the energy coordinate is shown in Fig. 1.8 (Harris 1979). Note the point indicated by ‡ which corresponds to the saddle point (transition state) on the potential energy surface.

The potential energy surface for the cyanohydrin formation reaction 1.27 has been calculated using a model requiring equilibrium constants and distortion energies, with allowance made for desolvation of CN^- prior to bond formation with the electrophilic centre (Guthrie 1996, 1998).

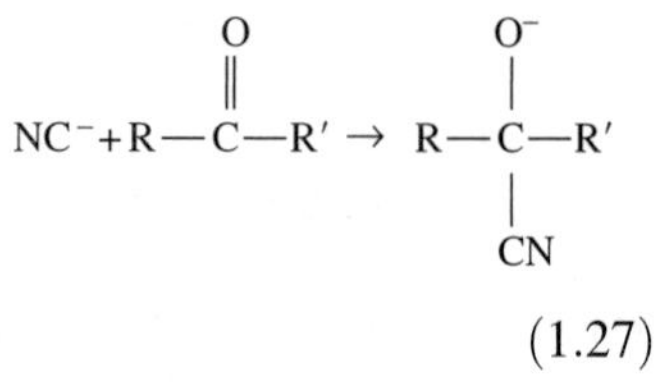

The idea of an 'imbalance' between nucleophile desolvation and bond formation in the transition state has been considered by Jencks (1987) and by Bernasconi (1992). Such an imbalance, or decoupling, can have a profound effect on reactivity, for example in terms from Bronsted type behaviour (Terrier *et al.* 1991, 1995). Interestingly this decoupling can be modulated by changing the solvent (DMSO-H_2O) composition (Buncel *et al.* 2002a; see also Chapter 7).

The free energy of solvation is a composite of the enthalpy and entropy. Entropy of solvation can also have large effects. Strong solvation usually implies loss of entropy, owing to relative immobilization of solvent molecules.

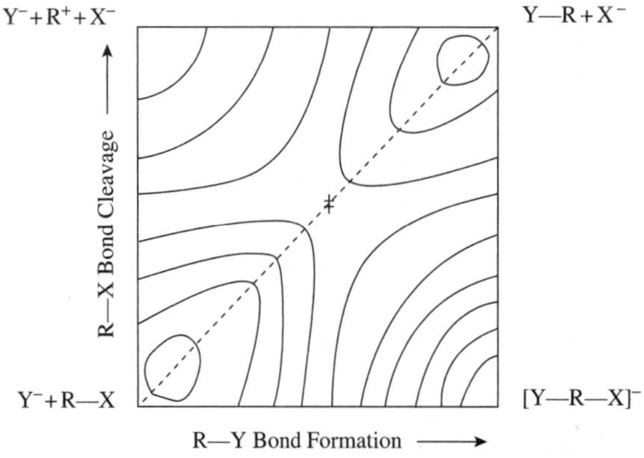

Fig. 1.8 Two-dimensional potential energy surface drawn as a contour map (After Harris *et al.* 1979, with permission).

Strong solvation of the reactant, therefore, makes the entropy of activation more positive, so (from eqn 2.20) makes the Gibbs energy of activation less positive, and the reaction therefore faster. The effect of strong solvation on the entropy of the activated complex, on the other hand, retards the reaction. Thus, the enthalpy and entropy of solvation of either the reactant or the activated complex have opposite effects. Prediction of the overall effect requires that these be disentangled. The required information concerning reactants is in principle available. That for activated complexes is not, though estimates may be made if data on molecules that resemble a postulated activated complex are known. In favourable cases, the dissection of the kinetics into these parts can be done (Blandamer 1977; Buncel and Wilson 1977, 1979; Buncel and Symons 1981; Tobe and Burgess 1999: 346, 363; Guthrie and Guo 1996).

The overall effect on the reaction rate thus depends on the free energies of the initial and transition states. The various possibilities, in terms of the free energy, are summarized qualitatively in Table 6.5. Reinforcement occurs if the transfer free energies of reactants and transition state have opposite signs. If they have the same sign, partial or complete balancing is expected.

For reactions in solution an additional thermodynamic property that can be helpful is available. The effect of pressure on the equilibrium constant of a reaction yields the volume change of reaction, ΔV, given by eqn 1.28.

$$\Delta V = -RT\left(\frac{\partial(\ln k)}{\partial p}\right)_T \tag{1.28}$$

The analogous effect of pressure on the rate constant gives the volume of activation, $\Delta_{\ddagger}V$, through eqn 1.29. Measurement of reaction rates at high pressures, as Tobe and Burgess (1999: 10) point out, requires specialized apparatus; nevertheless, a great many volumes of activation are now available: see Isaacs (1984), van Eldik *et al.* (1997, 2000), Blandamer and Burgess (1982), Laidler (1987), Tobe and Burgess (1999), and the large compilations of volumes of reaction and activation in the reviews by Drljaca *et al.* (1998 and references therein).

$$\Delta_{\ddagger}V = -RT\left(\frac{\partial \ln k}{\partial P}\right)_T \tag{1.29}$$

It is not likely that $\Delta_{\ddagger}V$ is large, though values outside $\pm 10\,\mathrm{mL\,mol^{-1}}$ have been obtained, notably for reactions consuming or generating ions in polar solvents. Van Eldik *et al.* (2000) show that in favourable cases there is a linear correlation between $\Delta_{\ddagger}V$ and ΔV. They present the example of substitutions on $Pd(H_2O)_4^{2+}$ in a variety of solvents, where $\Delta_{\ddagger}V \approx \Delta V - 2\ \mathrm{cm^3\,mol^{-1}}$. Where the activated complex resembles the products, this correlation is not unexpected, but it is by no means universal. Tobe and Burgess (1999) present *volume profiles*, which are schematic graphs of the volume changes along the reaction pathway, showing cases in which a degree of correlation exists (p. 536) and others in which it clearly does not (pp. 11, 301).

Volumes and entropies of activation for many classes of reactions show parallel trends, and can be interpreted in similar terms. In some cases the volume of activation is more reliable than the entropy of activation, because the latter is obtained by what is usually a long extrapolation of the plot of ln$k$ against $1/T$ to obtain the intercept. The volume of activation for a reaction in an inert solvent can be a help in the assignment of a mechanism, because dissociative activation may be assumed to result in a positive volume of activation (in the region of 10–15 $cm^3\,mol^{-1}$ for each bond assumed to be stretching in the activation process), and associative activation the reverse. In a solvent that interacts strongly with solutes, however, these interactions must also be taken into account. Reactions in which ions are generated, such as Menschutkin reactions (e.g. 1.30), are characterized in solution by large, negative, and solvent-dependent entropies and volumes of activation: $\Delta V^{\ddagger} = -12$ to -58 $cm^3\,mol^{-1}$ (Tobe and Burgess 1999),

$$Et_3N + Et\text{–}I \rightarrow [Et_3N^{\delta+} \cdots Et \cdots I^{\delta-}]^{\ddagger} \rightarrow Et_4N^+ + I^- \qquad (1.30)$$

because solvation of the nascent ions leads to reduced solvent freedom, reducing the entropy, and at the same time electrostriction of the solvent, reducing the volume, sufficiently to reverse the expected positive values.

Problems

1.1. Data are tabulated for the equilibrium between N,N'-bis-(hydroxymethyl)-uracil, **A**, and methanol, **B**, to form the di-ether, **C**, and water:

$$\mathbf{A} + 2\mathbf{B} \rightleftharpoons \mathbf{C} + 2H_2O$$

(Reagents were mixed in stoichiometric amounts.)

C_0	**0.5**	**1**	**2**	**3**	**4**
f at 17 °C	0.18	0.35	0.53	0.62	0.68
f at 30 °C	0.23	0.4	0.575	0.66	0.72

C_0 = initial concentration of **A**; f = fraction converted at equilibrium

(a) Calculate the mean equilibrium constant at each temperature.
(b) Calculate ΔG^0 at each temperature, and assuming they are constant, ΔH^0 and ΔS^0.
(c) With the same assumption, calculate the equilibrium constant and the fraction converted at 100 °C, $C_0 = 1.0$. (Hint: Try successive approximation.)

1.2. (a) Use the values of enthalpy of formation and entropies given below to calculate the equilibrium constant at 25 °C for the esterification reaction in the vapour phase:

$$CH_3COOH + C_2H_5OH = CH_3COOC_2H_5 + H_2O$$

Does it make any difference to the numerical value whether the constant is expressed in mole fraction, concentration or pressure units?

(b) Use the vapour pressures of the pure substances given to calculate the equilibrium constant (in mole fraction terms) for this reaction in a solvent in which all four substances form ideal solutions (a practical impossibility).

Substance (all as gas)	$\Delta_f H^0_{298}$ **(kJ mol^{-1})**	S^0_{298} **(J K^{-1} mol^{-1})**	p^0_{298} (mmHg)
Acetic acid	−434.3	282.7	15.4
Ethanol	−235.37	282	57.2
Ethyl acetate	−437.9	379.6	90.5
Water	−241.8	188.83	23.8

1.3. In dilute solution in 1,4-dioxane, the rational activity coefficients (as logarithms) of the four participants in the reaction in Question 2 are estimated as:

	Ethanol	**Acetic acid**	**Ethyl acetate**	**Water**
$\ln(f)$	1.02	1.39	0.08	2.1

Calculate the practical equilibrium constant of the same reaction in dioxane.

1.4. Pure p-xylene and water were equilibrated at 25 °C. The absorbance A_o of the aqueous layer measured in a 1-cm cell at $\lambda_{max} = 274$ nm (due to p-xylene) was 0.884. A solution of p-xylene, mole fraction $x_1 = 0.686$, and n-dodecane, similarly treated, gave absorbance $A = 0.749$. Assuming that both Beer's and Henry's laws hold for p-xylene in water and that n-dodecane is insoluble in water, what was the activity coefficient of p-xylene in the solution with n-dodecane?

1.5. The rate constants of the bimolecular reaction between OH radical and H_2 in the gas phase at 25, 45, and 100 °C were found to be 3.47×10^3, 1.01×10^4 and 1.05×10^5 L mol^{-1} s^{-1}, respectively. What are the Arrhenius activation energy E_a, the pre-exponential factor PZ, and the activation parameters $\Delta_{\ddagger}H$ and $\Delta_{\ddagger}S$? If the collision diameters of OH and H_2 are 310 and 250 pm, calculate Z and obtain an estimate of P, at 25 °C.

2 The solvent as medium

2.1 Intermolecular potentials

In this chapter we consider those aspects of the interaction of solvent and solute that are most clearly 'physical' in nature, setting aside the more 'chemical' aspects until the next chapter. The intermolecular potential characteristic of non-polar substances, the London or dispersion interaction, arises from the mutual, time-dependent polarization of the molecules. For two molecules well separated *in vacuo* it is approximated by the London formula:

$$V = -\frac{\alpha_1' \alpha_2' I_h}{3r^6} \tag{2.1}$$

where the factors α_1', α_2' are the polarizability volumes of the molecules, I_h the harmonic mean of their ionization energies and r is their separation. Equation 2.1 obviously is not accurate for a liquid, but it gives a sufficiently close estimate to enable Atkins (1998: 665) to conclude that the dispersion interaction is the dominant one in all liquids where hydrogen bonding is absent.

The remaining interactions in molecules that have permanent dipole moments, that is, the *dipole–dipole* and *dipole–induced-dipole* interactions, have the same dependence on intermolecular separation, varying as r^{-6}, and are of lesser magnitude at ordinary temperatures (Atkins 1998). These two interactions and the previously-mentioned dispersion interaction are collectively known as *van der Waals interactions*. They are related to such measurable properties as surface tension and energy of vaporization, and to concepts such as the internal pressure, the *cohesive energy density* (energy of vaporization per unit volume, $\Delta_{vap}U/V$) and the *solubility parameter*, δ, which is the square root of the cohesive energy density (Hildebrand and Scott 1962).

The *polarity* of the molecules is usually considered to be measured on a gross scale by the dielectric constant, and on a molecular scale by the electrical dipole and higher moments. Molecules lacking a dipole moment (e.g. carbon dioxide) may still exert short-range effects due to quadrupole, etc., moments. Dipolar bonds that are well-separated in a molecule seem to act almost independently on neighbouring molecules; Hildebrand and Carter (1930) showed that the three isomeric dinitrobenzenes, in their binary solutions in benzene, exhibit nearly identical deviations from Raoult's Law, though their dipole moments are different. The part of the electrical influence of a solvent on solute molecules that arises from the polarizability of the solvent molecules may be represented by the refractive index, n, or by functions of n such as the *volume polarization*, R, given by:

$$R = \frac{n^2 - 1}{n^2 + 2} \tag{2.2}$$

([R], the molar polarization, is R multiplied by the molar volume.) To allow for distortion polarization the refractive index should be that for far-infra-red radiation, but this is not usually known, so the usual sodium D-line value is commonly used. The corresponding quantity including the effects of the permanent dipole moment is of the form $P = (\epsilon_r - 1)/(\epsilon_r + 2)$, where ϵ_r is the relative permittivity, or dielectric constant. A possible measure of polarity, as distinct from polarizability, may then be defined by $Q = P - R$, that is by eqn 2.3:

$$Q = \frac{\epsilon_r - 1}{\epsilon_r + 2} - \frac{n^2 - 1}{n^2 + 2} \tag{2.3}$$

Sometimes, as when hydrogen bonding is possible, a more detailed charge distribution is important. This, however, is approaching the region of specific chemical effects, so will be deferred to the next chapter. Considerations of polarity become paramount when a solute is an ionic substance. We will consider first the category of relatively non-polar non-electrolyte solutions, then polarity, and finally, solutions of electrolytes in molecular solvents.

2.2 Activity and equilibrium in non-electrolyte solutions

Let us consider a general reversible reaction in solution:

$$\mathrm{A(solv)} + \mathrm{B(solv)} \rightleftharpoons \mathrm{C(solv)} + \mathrm{D(solv)} + \mathrm{etc.}$$

('A(solv)' represents a molecule of **A** surrounded by, but not necessarily strongly interacting with, molecules of the solvent). One may rigorously write down the expression for the thermodynamic equilibrium constant:

$$K = \frac{a_\mathrm{C} a_\mathrm{D}}{a_\mathrm{A} a_\mathrm{B}} \tag{2.4}$$

where the a's are activities. If the solution is ideal for all components, a highly unlikely event, all the activities may be replaced by mole fractions x:

$$K_x = \frac{x_\mathrm{C} x_\mathrm{D}}{x_\mathrm{A} x_\mathrm{B}} \tag{2.5}$$

This generally untrue statement can be rehabilitated by multiplying the mole fractions by *rational activity coefficients*, f, to reconvert them to activities:

$$K = \frac{x_\mathrm{C} f_\mathrm{C} x_\mathrm{D} f_\mathrm{D}}{x_\mathrm{A} f_\mathrm{A} x_\mathrm{B} f_\mathrm{B}} = K_x \frac{f_\mathrm{C} f_\mathrm{D}}{f_\mathrm{A} f_\mathrm{B}} = K_x \Gamma_f \tag{2.6}$$

so that if we can calculate or measure the activity coefficients, we can then use the activity-coefficient quotient Γ_f to predict the effect of the solvent on the value of the *practical equilibrium constant* K_x.

There is a certain freedom of choice still open to us as to the choice of the conditions in which the *true* or *thermodynamic equilibrium constant* is defined. Commonly, in working with solutions in a single solvent, one uses a definition according to which all activity coefficients become unity in the limit of extreme dilution. This is very convenient in that the practical and thermodynamic equilibrium constants become in the limit identical, and may not differ too much at moderate dilution. It will not do here, though, for we wish to focus attention on the changes that result from a change of solvent, even if the solutions in both solvents are exceedingly dilute. We therefore must choose a standard state for each solute that is the same regardless of the solvent, and hence a single K that is a function of temperature alone. The limit that each practical constant, K_x, K_P, or K_C, approaches as the concentrations are decreased will still depend on the properties of the solvent. The simplest choice for this single K is that for the reaction in the gas phase at low pressure (so the Ideal Gas Law applies to all species), with no solvent present:

$$A(g) + B(g) \rightleftharpoons C(g) + D(g): \quad K = K_P$$

We now imagine a solvent in which all the solutes form ideal solutions, and applying Raoult's Law (eqn 1.5) to each p in K_P (eqn 1.4 rewritten for this reaction, with $\Gamma_\gamma = 1$), we obtain:

$$K_P = \frac{P^0_C x_C P^0_D x_D}{P^0_A x_A P^0_B x_B} = K_x(\text{ideal})\frac{P^0_C P^0_D}{P^0_A P^0_B} \tag{2.7}$$

Since the P^0s, the vapour pressures of the several pure substances, depend only on temperature, K_x(ideal) is a true constant, just as good as K_P. This is probably the most convenient candidate for the thermodynamic constant, K, in all cases of interest to us except those involving ionic species. Unless a comment is made to the contrary, it may be assumed that $K = K_x$(ideal) in the following pages.

Deviations of K_x from K can be of three sorts. Deviations at low and moderate concentrations from the zero-concentration limiting value are most important where ionic solutes are dissolved in non-ionic solvents. Even at extreme dilution, departures from Raoult's Law on the part of each solute can arise in two distinct ways, with opposite effects. Deviations that are due to specific chemical or quasi-chemical attractive interactions between unlike molecules, and that lead to enhanced mutual solubilities, lower partial vapour pressures, and activity coefficients less than unity are called *negative deviations*. Those that arise from mere differences between the molecules of the two kinds, such as differences of size or shape or of the intensity of intermolecular forces (reflected in differences in the solubility parameter, mentioned above), and that lead to diminished solubility, higher partial vapour pressures, and activity coefficients greater than unity, are called *positive deviations* (see Fig. 1.2).

The effect of mere difference between molecules, such as different size, shape or polarizability, as it affects intermolecular forces was considered by Hildebrand and Scott (1962). They were led to the concepts of *cohesive energy density* (which is defined as the molar energy of evaporation divided by the molar volume, and which has the dimensions of pressure) and its square root, the *solubility parameter*, defined by:

$$\delta = (\Delta U_{\text{vap}}/V_{\text{mol}})^{1/2} \tag{2.8}$$

When two liquids are mixed, if the molecules distribute themselves randomly (thermal agitation being enough to overcome any tendency to specific pairing or segregation), the resulting solution is termed 'regular'. Assuming regular solution, and neglecting any volume changes, Hildebrand and Scott derive the relation:

$$RT \ln a_2 = RT \ln x_2 + V_2\phi_1^2(\delta_2 - \delta_1)^2 \tag{2.9}$$

in which a_2, x_2, V_2, and δ_2 are the activity, the mole fraction, the molar volume, and the solubility parameter of component **2** in the mixture, and ϕ_1 and δ_1 are the volume fraction and the solubility parameter of component **1**. If the solution were ideal a_2 would be equal to x_2, so the last term is a measure of the departure of the solution from ideality. Note that in pure **2**, where ϕ_1 is zero, $a_2 = x_2 = 1$; in all very dilute solutions the solvent behaves as if the solution were ideal. In terms of the rational activity coefficient, f_2:

$$RT \ln f_2 = V_2\phi_1^2(\delta_2 - \delta_1)^2 \tag{2.10}$$

In a very dilute solution of **2** in **1**, on the other hand, $\phi_1 \approx 1$, so f_2 is constant, and the solute obeys Henry's Law. Very great approximations were used in the derivation of these equations, but they are surprisingly good as long as the components of the mixture do not carry too large dipole moments or take part in hydrogen bonding.

Now let us consider a reaction among four species **A, B, C,** and **D** in dilute solution in a solvent **S**:

$$\mathbf{A} + \mathbf{B} \rightleftharpoons \mathbf{C} + \mathbf{D}$$

We will use eqn 2.10 four times, putting **A, B,** etc. in turn as component **2**, and **S** as component **1**, obtaining eqn 2.11, and we may write the practical equilibrium constant: $K_x = K/\Gamma_f$.

$$\begin{aligned}\ln\frac{f_C f_D}{f_A f_B} &= \frac{\phi_S^2}{RT}[V_C(\delta_C - \delta_S)^2 + V_D(\delta_D - \delta_S)^2 - V_A(\delta_A - \delta_S)^2 - V_B(\delta_B - \delta_S)^2] \\ &= \frac{\phi_S^2}{RT}\Delta[V(\delta - \delta_S)^2] \\ &= \ln \Gamma_f \end{aligned} \tag{2.11}$$

Since the thermodynamic constant K is a good constant, depending only on the temperature, anything that decreases Γ_f will increase K_x, and will increase the relative amounts of **C** and **D** present at equilibrium. This will be the case if the $(\delta - \delta_S)^2$ terms are kept small for **C** and **D**, and large for **A** and **B**. That is to say, if we want a good yield of **C** or **D**, we should carry out the reaction in a solvent as much like **C** or **D**, or both, as possible, and unlike **A** and **B**, in the sense of the saying, 'Like dissolves like'.

For example, consider the reaction:

$$CO + Br_2 \rightleftharpoons COBr_2$$

(a somewhat artificial example, for the equilibrium lies too far to the right for convenient measurement in solution at room temperature, but it illustrates the method in a case where the result is not obvious without calculation). The relevant quantities, assuming the partial molal volumes of the solutes are equal to their molal volumes in the pure liquids, are:

	CO	Br_2	$COBr_2$
δ (MPa$^{1/2}$)	13.5	23.7	17.4
V (mL mol^{-1})	35	55	92

Equation 2.11 with these values leads to:

$$\ln(K_x/K) = 3.80 - 0.1414\delta_S - 0.0008\delta_S^2 \tag{2.12}$$

It would have been hard to guess whether the increase in δ from CO to the product, or the decrease from Br_2 to the product, would dominate the effect. As a rough estimate, where the volume change of the reaction is small (here $\Delta V = 92 - 35 - 55 = 2$ mL) the slope of the plot of $\ln(K_x/K)$ versus δ_S is approximately equal to $2\Delta(V\delta)/RT$, here about -0.14. The negative slope means that the formation of the product is favoured in a medium of low solubility parameter.

2.3 Kinetic solvent effects

The above approach may be applied to the effect of solvent on the rate of a reaction, through the thermodynamic interpretation of the Transition State Theory (see Chapter 1). Representing a reaction as:

$$\mathbf{A} + \mathbf{B} \rightleftharpoons \ddagger \rightarrow \text{products}$$

we apply eqn 2.10 to the activated complex ‡ and to the reactants, as before, and obtain:

$$\begin{aligned}\ln\frac{k}{k_0} &= \ln\frac{1}{\Gamma_f} = \ln\frac{f_A f_B}{f_\ddagger} \\ &= \frac{\phi_S^2}{RT}[V_A(\delta_A - \delta_S)^2 + V_B(\delta_B - \delta_S)^2 - V_\ddagger(\delta_\ddagger - \delta_S)^2] \\ &= \frac{\phi_S^2}{RT}[(V_A\delta_A^2 + V_B\delta_B^2 - V_\ddagger\delta_\ddagger^2) - 2(V_A\delta_A + V_B\delta_B - V_\ddagger\delta_\ddagger)\delta_S \\ &\quad + (V_A + V_B - V_\ddagger)\delta_S^2] \\ &= -\frac{\phi_S^2}{RT}[\Delta_\ddagger(V\delta^2) - 2\Delta_\ddagger(V\delta)\delta_S + \Delta_\ddagger(V\delta_S^2)]\end{aligned} \tag{2.13}$$

The molar volumes of the reactants, V_A, V_B, are presumably known. The δ's are known for many substances or can be calculated from known quantities

(Hildebrand and Scott 1962; Marcus 1985). The molar volume of the activated complex is not usually known; $\Delta_{\ddagger}V$ may be obtained through the effect of pressure on the reaction rate (see Section 1.10). Finally, $\delta_{\ddagger}$ may be obtained from the best fit of data for the rate constant of the reaction in a variety of solvents to eqn 2.13.

An example occurred in a study of the rate of oxidation of toluene by chromyl chloride (Étard 1881), in assorted solvents (Stairs 1962):

$$C_6H_5CH_3 + CrO_2Cl_2 \rightarrow C_6H_5CH_2OCrOHCl_2 \qquad k_1 \text{ (slow)}$$
$$C_6H_5CH_2OCrOHCl_2 + CrO_2Cl_2 \rightarrow C_6H_5CH(OCrOHCl_2)_2 \quad k_2 \text{ (somewhat faster)}$$

The structures of the intermediate and of the final product, a brown precipitate insoluble in all solvents that do not destroy it, are conjectural. The latter appears to contain Cr(IV). On hydrolysis, benzaldehyde, HCl and a mixture of Cr(III) and Cr(VI) species, with some benzyl chloride are formed.

In Fig. 2.1(a) the logarithm of the observed rate constant k_1 is plotted against the solubility parameters of the six solvents. The curve was drawn with the assumed value $-10\,\text{cm}^3$ for $\Delta_{\ddagger}V$, and fitted to all the points but one (more about that one below). The moderate increase in rate with increase in δ_S was taken as evidence that the activated complex had a higher solubility parameter

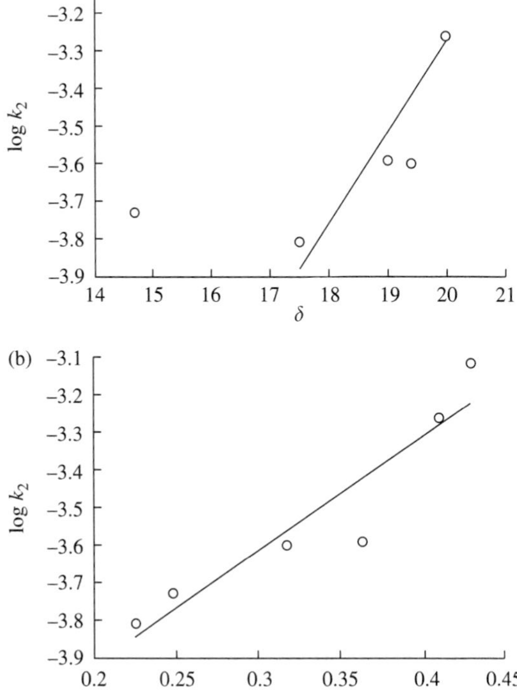

Fig. 2.1 Kinetic solvent effect in Étard's reaction: (a) $\text{Log}_{10}(k)$ versus Hildebrand's solubility parameter. (b) $\text{Log}_{10}(k)$ versus Kirkwood's dielectric function. Redrawn from Stairs (1962).

than the reactants, but not much higher, and that it was probably not ionic in nature. This was helpful in an attempt to assign a mechanism to the reaction.

2.4 Polarity

The preceding treatment was based on the assumption that none of the molecules involved is so polar as to exert strong orienting forces or specific attractive forces on neighbouring molecules. It breaks down in cases such as propanone (acetone), which has a rather low solubility parameter based on its energy of vaporization, but which must be assigned a high and variable one to account for its ability to dissolve liquids such as water, and even inorganic salts. It dissolves these substances through specific interactions: hydrogen bonding with water molecules, and ion–dipole solvation of the cations of salts. A less extreme case is apparent in the kinetic data discussed in the last section. The point off the curve in Fig. 2.1(a) is for the solvent 1,1,2-trichloro-1,2,2-trifluoroethane (Freon 113), which has a low solubility parameter ($\delta = 14.7\,\mathrm{MPa}^{1/2}$) owing to the weak van der Waals (London) forces typical of highly fluorinated molecules, but which is nevertheless somewhat polar ($\mu = 0.4$ Debye, approximately[1]). It has about the same effect on the rate as tetrachloromethane, which, though non-polar, has higher London forces, leading to $\delta = 17.9$.

It is possible to extend the solubility-parameter method to include the effects of moderate polarity by assuming the cohesive energy density $\Delta_{\mathrm{vap}}U/V$) to be made up of two parts, $\delta^2 + \omega^2$, where δ^2 is a measure of the London interaction and ω^2 (which is proportional to μ^4) of the polar interaction. With this complication it becomes less convenient, however, and less satisfactory.[2]

2.5 Electrostatic forces

A very different approach from the foregoing was made by Kirkwood 1934, Onsager 1936, and Amis and Hinton 1973: 241. Kirkwood looked at the molecules of the solute as spheres, each bearing at its centre an electric dipole moment μ, in a continuous medium of dielectric constant ϵ_r. (The dielectric constant, or relative permittivity, is the ratio of the permittivity ϵ of the medium to the permittivity of free space, ϵ_0.) The difference in free energy of a mole of such spherical dipoles in this medium from what it would be if the dielectric constant were unity is given by:

$$\begin{aligned} G_{\epsilon_r} - G_1 &= RT\ln f \\ &= -\frac{L\mu^2}{4\pi\epsilon_0 r^3}\frac{(\epsilon_r - 1)}{(2\epsilon_r + 1)} \end{aligned} \tag{2.14}$$

[1] The Debye unit of dipole moment is equal to 10^{-18} esu cm, or 3.336×10^{-30} C m. in SI.
[2] The reader may pursue the matter in treatments by Hildebrand and Scott (1950, Ch. IX) and by Burrell (1955), and in the review by Barton (1975).

Here f is the activity-coefficient, L is Avogadro's number and r is the radius of the sphere. Applying eqn 2.14 to each of the species in the reaction: **A** + **B** = **C** + **D**, we obtain:

$$RT \ln \Gamma_f = -\frac{N(\epsilon_r - 1)}{(2\epsilon_r + 1)4\pi\epsilon_0}\left(\frac{\mu_C^2}{r_C^3} + \frac{\mu_D^2}{r_D^3} - \frac{\mu_A^2}{r_A^3} - \frac{\mu_B^2}{r_B^3}\right) \tag{2.15}$$

Again recalling that $K_x = K/\Gamma_f$, to obtain a good yield of **C** or **D** we want Γ_f to be small. This turns out to lead to the same rule of thumb as in the non-polar treatment, that the best solvent is the one that most resembles the products. *If the desired products are more polar than the reactants, the reaction is favoured by a polar solvent, but if the products are less polar, a relatively non-polar solvent is preferred.*

Equation 2.15 may be applied to the reaction considered above, between bromine and carbon monoxide. Bromine is non-polar. Carbon monoxide has a small dipole moment, of magnitude 0.112 Debye, and the product, carbonyl bromide, has a somewhat larger dipole moment, about 1.2 Debye. The mean radii of the three molecules approximated as spheres are: (Br_2) 0.23 nm, (CO) 0.20 nm, ($COBr_2$) 0.296 nm. If these values are used in eqn 2.16, the result is:

$$\log_{10}\left(\frac{K_x}{K}\right) = 0.566\frac{\epsilon_r - 1}{2\epsilon_r + 1} \tag{2.16}$$

As would be expected, the formation of the more polar product is favoured in solvents of higher dielectric constant (contrary to the prediction based on the solubility parameters).

Solvents of low dielectric constant tend to also have low solubility parameters. Figure 2.2 shows a plot of the dielectric constant function against the solubility parameter for 26 solvents of varied character, including the six solvents used in the study of Étard's reaction, above. Since the correlation between these two properties is clearly weak, it should be possible in some cases to tell whether the polarity or the non-polar (London) part of the intermolecular forces have the more important part in determining the total effect of the solvent on the yield of product. For the reaction of carbon

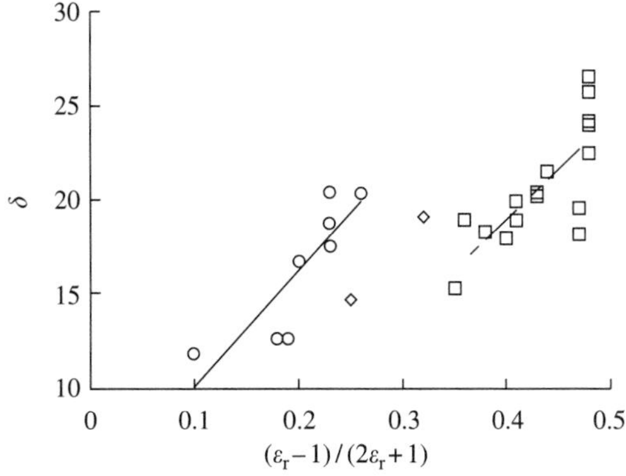

Fig. 2.2 Hildebrand's δ ($MPa^{1/2}$) plotted versus Kirkwood's dielectric function. Circles, non-dipolar (or nearly) solvents; squares, strongly dipolar; lozenges, intermediate.

monoxide with bromine, the two theories predict opposite effects, so it is unfortunate that this reaction is a difficult one to study.

As in the previous treatment, this method can also be applied to reaction rates, through the thermodynamic interpretation of Transition-State Theory. Figure 2.1(b) shows the result of applying eqn 2.15 to the same kinetic data for Étard's reaction, discussed above in terms of solubility parameters. Here the rate constants are plotted (as common logarithms) against $(\epsilon_r - 1)/(2\epsilon_r + 1)$. Comparing Fig. 2.1(b) to Fig. 2.1(a), it is immediately apparent that the point for Freon, which is well off the curve in (a), is close to the line in (b), suggesting that the electrostatic part of the effect is more important, though the parallel trends in the solubility parameters and the dielectric functions of the rest of the solvents make it difficult to confirm this conclusion (it was difficult to find solvents with which chromyl chloride did not react). The dipole moment of the activated complex may be estimated from the slope of the line in Fig. 2.1(b) as about 2.0 Debye, which is a value typical of rather polar molecules.

As another example, Reichardt (1988: 154) cites data from Huisgen and co-workers (Swieton *et al.* 1983) for the cycloaddition of diphenylketene to n-butyl vinyl ether. Figure 2.3 shows a plot of the natural logarithm of the rate constant (relative to the slowest) versus the Kirkwood function. The slope corresponds to a dipole moment for the transition state of about 10 Debye (34×10^{-30} C m), larger than that of the product, indicating a considerable degree of charge separation.

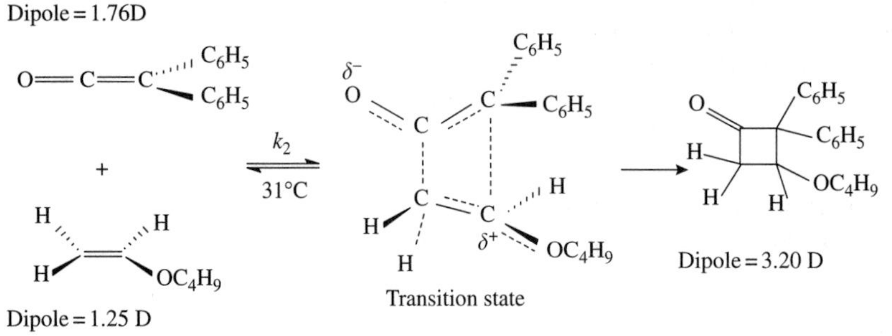

Dipole = 1.76D

$O{=}C{=}C(C_6H_5)_2$ + $H_2C{=}CH{-}OC_4H_9$ $\underset{31°C}{\overset{k_2}{\rightleftharpoons}}$ Transition state $\longrightarrow$ cyclobutanone product

Dipole = 1.25 D

Transition state (δ^- on O, δ^+ on C)

Dipole = 3.20 D

2.6 Electrolytes in solution

When a typical salt is dissolved in a liquid, such as water, in which it is a strong electrolyte, the ions of the salt interact both with the solvent molecules and with each other. The latter interaction persists on dilution to very low concentrations, for the Coulomb force between like or oppositely charged ions extends to long distances. It depends on concentration in a way that can be calculated, at least in dilute solutions, by the Debye–Hückel (1923) Theory. This theory has been fully treated elsewhere (e.g. Barrow 1988; Skoog *et al.* 1989; Atkins 1998; 248–253), so let us note here its main features.

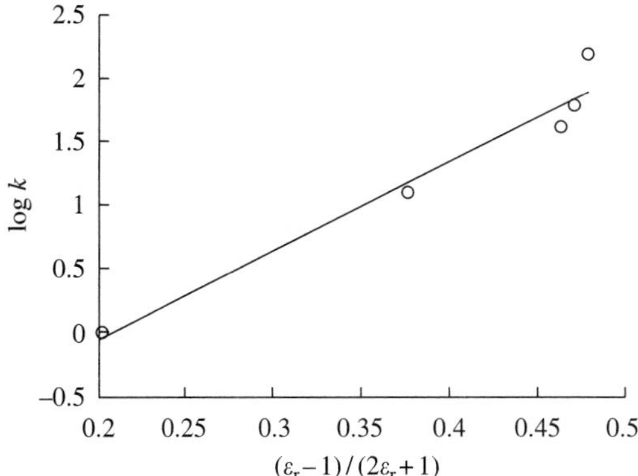

Fig. 2.3 Cycloaddition of diphenylketene to butyl vinyl ether. Logarithm of the relative rate constant versus Kirkwood's dielectric function. Data from Reichardt (1988: 154).

The theory is commonly quoted at several levels of approximation. The simplest form, which is valid only at the lowest concentrations (below $0.001\,\mathrm{mol\,L^{-1}}$ in water, and lower in most other solvents), is the *limiting law* (for a solution of a single salt):

$$\log_{10} \gamma_{\pm} = Az_{+}z_{-}c^{1/2} \tag{2.17}$$

The constant A depends on the dielectric constant of the solvent and on the temperature, in the form: $1.8246 \times 10^{6}(\epsilon_r T)^{3/2}$. For water at 25 °C its value is $0.5115\,\mathrm{L^{1/2}\,mol^{-1/2}}$. The *mean ionic activity coefficient*, $\gamma_{\pm}$, is defined so that the activity $a = \gamma_{\pm}c$ becomes equal to the molar concentration c in the low concentration limit. The subscript $\pm$ is added to the symbol because it is not possible rigorously to define the activities of the separate ions, nor their activity coefficients. By assuming that the activity coefficients of the two ions of a symmetrical electrolyte are equal, however, it is possible to write a single-ion version:

$$\log_{10} \gamma = -Az^{2}c^{1/2} \tag{2.18}$$

The negative sign appears because one of the zs in eqn 2.17 is negative.

For a solution containing more than two kinds of ions it is necessary to define the *ionic strength*, I, by the relation:

$$I = \frac{1}{2}\sum_{i} z_i^{2} c_i \tag{2.19}$$

With this definition, the limiting law for a single ion becomes:

$$\log_{10} \gamma = -Az^{2}I^{1/2} \tag{2.20}$$

To extend the application of the theory to more useful concentrations, the fact that two oppositely-charged ions can approach each other only until their centres are separated by the sum of their radii, a, is used to correct eqn 2.20 to read:

$$\log_{10} \gamma = -\frac{Az^{2}I^{1/2}}{1 + BaI^{1/2}} \tag{2.21}$$

The value of the new constant B in water at 25 °C is $3.291 \times 10^9\,\mathrm{m^{-1}\,mol^{-1/2}\,L^{1/2}}$. For many electrolytes the value of a is around 300–400 pm, so the product Ba is close to unity for aqueous solutions at room temperature. Taking advantage of this coincidence, and adding an empirical linear term in I (which may be justified by some qualitative reasoning), Davies (1962) has formulated a useful approximate expression, eqn 2.22, which he finds applicable to a great many ionic species in water at room temperature.

$$\log_{10}\gamma_z = -0.5z^2\left(\frac{I^{1/2}}{1+I^{1/2}} - 0.3I\right) \tag{2.22}$$

Figure 2.4 illustrates the form of the limiting law (eqn 2.20), the 'extended' eqn (2.21) and Davies's approximation (2.22), all for a univalent ion in aqueous solution at 25 °C.

At low and moderate concentrations, the theory predicts that the logarithm of the activity coefficient will be negative, and inversely dependent on the product of the temperature and the dielectric constant of the solvent. At the lowest concentrations the dependence is on $(\epsilon_0 T)^{-3/2}$, but at moderate concentrations the exponent approaches -1. The effect of interionic forces at moderate ionic strengths is to favour the formation of ionic products, for $\gamma_\pm$ in this region is always less than unity, but this effect is diminished as the dielectric constant increases. This diminution appears to contradict our expectation that a more polar solvent would favour the formation of ionic products, for ionization may be viewed as polarity carried to the extreme. It is, however, a small correction on the larger interaction of the ions with the solvent, discussed below.

As the concentration is increased further, deviations from the simpler forms of the Debye–Hückel Theory accumulate. The concentration beyond which eqn 2.21 fails depends strongly on the quantity ϵT. In water at room temperature, where ϵT is about 24000, the useful limit is in the range 0.01–0.1 mol L^{-1} for uni-univalent salts, lower for higher valences. In liquid ammonia at its boiling point, -33 °C, $\epsilon T \simeq 4800$, it is less than 0.001 mol L^{-1}. Furthermore, in solvents of low dielectric constant the ions tend to associate in pairs or higher aggregates (Section 2.9), so that in many solvents no strong

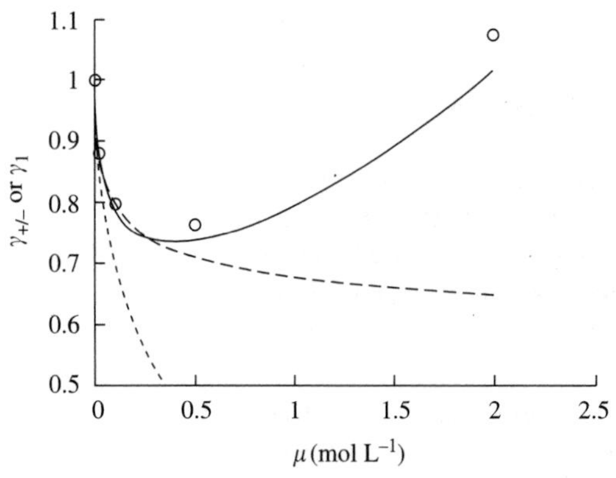

Fig. 2.4 The Debye–Hückel limiting law (eqn 2.20) fine dashed curve, the corrected law (eqn 2.21) coarse dashed curve, Davies's approximation (eqn 2.22) solid curve for $z = 1$, and experimental data for HCl, all in water at 25 °C. The activity coefficient (as common logarithm) plotted against the ionic strength, $\mu = \sum z_i^2 c_i/2$.

electrolytes exist. Nevertheless the theory is accurate over a limited range of concentrations, and (in the form of Davies's approximation) a useful guide over a wider one. It provides a mathematical form to aid extrapolation of certain data to very low concentrations, where the effects of interionic forces may become negligible, and the ion–solvent interactions may be separately examined.

2.7 Solvation

When the ions are so diluted that their mutual influences may be neglected, there still remains their interaction with the solvent, which is termed 'solvation' (not 'salvation', as a Calgary newspaper once headlined a report of a conference on this topic).[3] Solvent molecules may interact so strongly with a dissolved ion that they become firmly bound by ion–dipole forces, or by covalent bonds, and for certain purposes may be counted as part of the ion. Attempts have been made by various means to determine the number of molecules so bound, the *solvation number*. Different methods give different results, which should not be surprising, for the meaning of 'firmly bound' depends on how hard one tries to dislodge them. Solvation numbers of ions in various solvents as found by different methods are discussed by Hinton and Amis (1971). Methods that have been used include measurement of apparent hydrodynamic radii of ions in electrical conductance and of the amounts of one solvent (usually water) carried into another in an extraction process. The amounts of solvent incorporated in crystals as 'solvent of crystallization' have been used, though crystals may include solvent molecules not bound to any ion, but occupying sites elsewhere in the lattice. In $KAuBr_4 \cdot 2H_2O$, for example, the water molecules are not attached to any ion, but occupy vacant spaces in the lattice (Cox and Webster 1936). Various methods tend to suggest that sodium chloride in aqueous solution, for instance, carries between 4 and 8 water molecules about the Na^+ ion, and fewer about the Cl^- ion. Because the bound molecules are fully polarized, in the sense of being aligned with the ionic field, the inner sphere they occupy has been called the *sphere of dielectric saturation*.

The region about a solute particle (molecule or ion) within which the structure of the solvent is altered from what it is in the pure solvent, or in solution remote from a solute particle, is called the *cybotactic region*. (see Fig. 2.5) About an ion one may distinguish the inner or coordination sphere, within one solvent molecular diameter, where solvent molecules are more or less firmly bound and strongly oriented, depending on the charge-to-size ratio of the ion and the polarity of the solvent molecules. Here the solvent structure is essentially destroyed, and the ion with its bound solvent acts as a larger ion. In the next region the solvent molecules, if dipolar, are

[3] On the other hand, E. S. Amis and J. F. Hinton (1973) dedicated their book, 'Solvent Effects on Chemical Phenomena', Vol. 1, Academic Press, New York and London, 'To Dr. E. A. Moelwyn-Hughes, retired general of the Solvation Army'.

oriented by the ionic field in diminishing degree as their distance from the ion increases. The conflict between this orientation and the tendency of the solvent to assume its normal structure may result in a more chaotic structure in this region, reflected, for instance, in diminished viscosity. Both the entropy of solution and the partial molar volume of the solute are decreased by the binding in the inner sphere, and increased by the disorder in the outer sphere. Which effect predominates depends on the size and charge of the ion.

The presence of a non-polar solute in a polar solvent creates a rather different situation, described, in discussing aqueous solutions, as the *hydrophobic effect*, illustrated in Fig. 2.5(b). Here the water molecules, unable to form hydrogen bonds with the solute, do so with each other, but in a way that creates a cavity within which the solute is contained. A quasi-crystalline inner sphere is formed, different in structure from the bulk water structure. Outside the inner sphere is again a chaotic region, before normal solvent structure is resumed at greater distances. Okazaki *et al.* (1979), in a Monte Carlo simulation (see Chapter 5) of methane in water, were able to distinguish these three regions. Both the enthalpy and the entropy are decreased, so the free energy change may have either sign. Abraham (1982) illustrated this by compiling data on the enthalpy, entropy, and free energy of solvation of 28 non-polar, gaseous substances in 32 solvents of varied character.

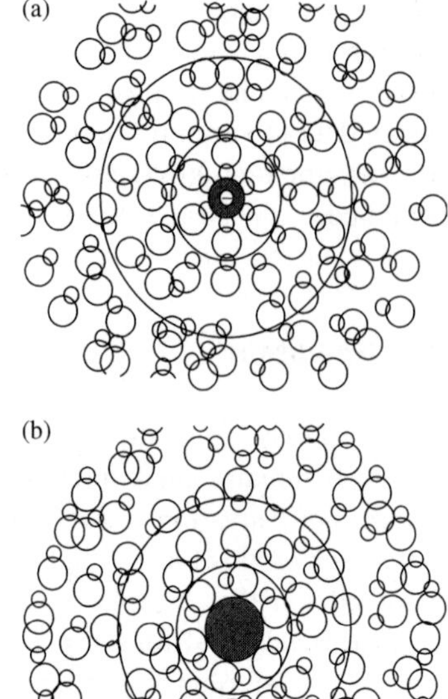

Fig. 2.5 (a) Solvation in Flatland. A solvated negative ion (shaded) and its cybotactic region. The solvent molecules within the innermost circle are virtually fixed in orientation toward the ion. Those within the next circle are less strongly oriented, while those beyond are undisturbed. (b) Hydrophobic solvation in Flatland: cybotactic region around a non-polar solute (shaded) in a polar solvent. Here the innermost solvent molecules are H-bonded to each other in a ring about the solute, with which they hardly interact. The region of disturbed solvent structure is within the outer circle.

When two molecules thus solvated approach one another closely, the two cavities may merge, forming a single, larger cavity, and liberating some of the solvent. Both the energy and the entropy will increase. The entropy increase usually dominates, so dimerization is favoured, that is, the equilibrium constant for the reaction depicted in Fig. 2.6 is greater than unity.

If the solute contains ions such as carboxylate, with a charged head and a non-polar tail, both ionic coordination and the hydrophobic effect can be simultaneously present. Soaps, such as sodium stearate $CH_3(CH_2)_{16}COO^-Na^+$, and detergents, such as sodium dodecylsulfate $CH_3(CH_2)_{11}OSO_3^-Na^+$, have long non-polar tails. In aqueous solution above a certain concentration, the *critical micelle concentration*, they form *micelles*, more-or-less spherical globules in which the tails are together in the interior, and the polar or ionic heads on the surface. Figure 2.7 illustrates a spherical micelle.

The hydrophobic effect can lead in the extreme case to the formation of crystalline hydrates, such as the well-known chlorine hydrate $Cl_2 \cdot 8H_2O$, and $CH_4 \cdot nH_2O$, found at great depth in ocean sediments, and recently of interest (Kleinberg and Brewer 2001) as a large potential source of energy. These are examples of the general class of substances called *clathrates*.

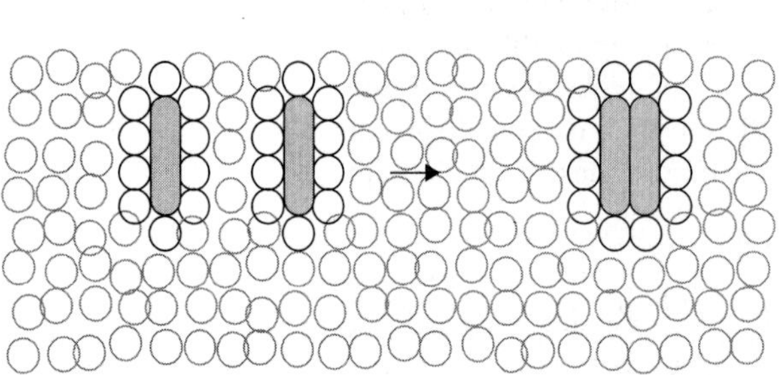

Fig. 2.6 Hydrophobic dimerization: $2M(solv) \rightarrow M_2(solv)$. Eight solvent molecules shown as being liberated.

2.8 Single ion solvation

It is not possible by thermodynamic arguments alone to decide how the volumes of the ions and their solvation spheres are to be partitioned between the cation and anion of a dissolved salt. The conventional solution of this problem, as far as aqueous solutions are concerned, has been to make an arbitrary assignment of zero to the volume of the hydrogen cation. This is obviously a fiction, because the proton in water is known to be associated with at least one water molecule to form the 'hydronium' or 'hydroxonium' ion, H_3O^+, or more, as $H_9O_4^+$, for example.

Nevertheless, the additivity of ionic properties in dilute solution ensures that all the relevant properties of the solute are correctly represented. Trémillon (1974) and Marcus (1985: 96–105) present discussions of the various ways that partitioning of this and other ionic properties may be made, using arguments outside thermodynamics. The large size and similar structure of tetraphenylarsonium (**1**) and tetraphenylborate (**2**) ions have led to suggestions that they should have the same size in most solvents, and be nearly equally solvated.

The conclusion Marcus comes to is that the best value for the partial molar volume of the hydrogen cation in extremely dilute aqueous solution is $-6.4\,\mathrm{cm^3\,mol^{-1}}$. The reason for this negative value is that by tightening the structure of water in its vicinity, through enhanced hydrogen bonding, and in more subtle ways (encompassed in the term *electrostriction*), the proton causes a local shrinkage. Similar effects are seen with many ions, especially those with small crystal radii or large charges. Other properties, such as enthalpies, Gibbs energies and entropies may be partitioned by similar arguments (Marcus 1985: 105–113; Cox 1973).

Outside the primary sphere is one in which the molecules are somewhat polarized, but not intensely so. If we take the radius of the ion r to be the radius of the primary sphere, we may follow Born (1920) and Hunt (1963) and calculate the free energy of transfer of an ion of radius r from one medium to another (let us assume that one is water and the other a solvent **S**) according to eqn 2.23, or for a mole of a 1 : 1 electrolyte, eqn 2.24:

$$\Delta G = -\frac{z^2e^2}{8\pi\epsilon_0}\left(\frac{1}{\epsilon_W}-\frac{1}{\epsilon_S}\right)\frac{1}{r} \tag{2.23}$$

$$\begin{aligned}\Delta G_{tr} &= -\frac{N_Ae^2}{8\pi\epsilon_0}\left(\frac{1}{\epsilon_W}-\frac{1}{\epsilon_S}\right)\left(\frac{1}{r_+}+\frac{1}{r_-}\right)\\ &= RT\Delta\ln\gamma_\pm^2\end{aligned} \tag{2.24}$$

Numerical calculation of these three effects, i.e. the *primary and secondary solvation* and the interionic interaction, unfortunately increases in difficulty in the order of increasing importance. As Hunt shows, following Latimer *et al.* (1939), one may allow for the difficult-to-calculate primary solvation for a series of similar ions in a single solvent by assigning an 'effective radius' and applying the Born equation. To approach this difficult situation experimentally, consider a reaction among a number of ionic or polar solutes in a medium consisting of a relatively non-polar solvent containing some water,

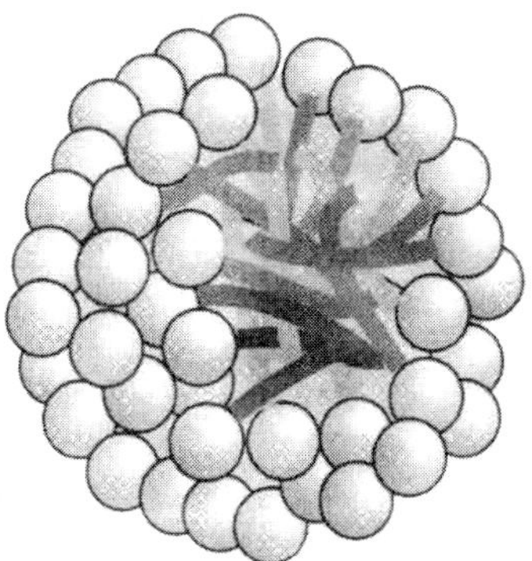

Fig. 2.7 A schematic version of a sperical micelle. The hydrophilic groups are representd by spheres and the hydrophobic hydrocarbon chains are represented by the stalks; these stalks are mobile (reproduced from Atkins (1994: 975) with permission).

1

2

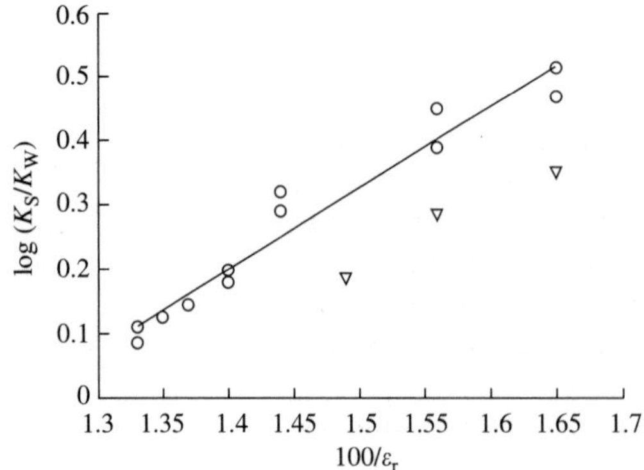

Fig. 2.8 Acids in aqueous/alkanol mixed solvents: $\log(K_S/K_W)$ versus $1/\varepsilon_r$. Redrawn from Robinson and Stokes (1959: 356), with permission. Points marked by triangles are not included in the statistics.

in which water molecules preferentially solvate the polar and ionic solutes. The primary solvation is by water and may be assumed to be constant. The remaining effects, attributed to the secondary solvation, may then be seen. Robinson and Stokes (1959) present a graph (redrawn as Fig. 2.8) in which the dissociation constants of a number of carboxylic acids in a number of aqueous mixed solvents are compared with those for the same acids in water. All the solvent mixtures contained enough water that the reaction in each case could be assumed to be:

$$HA + (x+y)H_2O \rightleftharpoons A^- \cdot xH_2O + H_3O^+ \cdot yH_2O.$$

The dependence of $\log_{10}(K_S/K_W)$ on the reciprocal of the dielectric constant was clear, though a few points fall off the line. By neglecting any effect on the activities of the neutral species, we use eqn 2.24 to calculate from the slope (about 133 ± 7) of the line in the figure, 200 ± 10 pm for the harmonic mean of the radii of hydrated H^+ and carboxylate. This seems somewhat small, but not impossible. The aberrant points may reflect failure of either of the assumptions: strongly preferential solvation by water or absence of medium effects on the unionized acid.

2.9 Ionic association

The Debye-Hückel Theory has been very successful in explaining the behaviour of strong electrolytes in solvents of relatively high dielectric constant, at low concentrations. The deviations noted above at higher concentrations are attributed to various short-range interactions between ions. In very concentrated solutions, the solution begins to be crowded, and begins to resemble a molten salt, with the beginnings of short-range order. This situation is not treatable by the methods contemplated here. At more moderate concentrations, oppositely charged ions may be significantly associated. In solvents with dielectric constants much less than that of water, hardly any electrolytes are strong (in the sense of being fully dissociated into separated, solvated ions). Thus in most solvents the study of electrolytes is mainly concerned with weak electrolytes. Quite aside from weak acids and bases, and the few salts, such as $HgCl_2$, that have genuine molecules, this nearly universal weakness in solvents of low dielectric constant needs explanation.

The problem has been approached by a number of workers, notably Bjerrum (1926), Gronwall *et al.* (1928) and Fuoss and Krauss (1933). These are discussed by Davies (1962). All agree that the dielectric constant of the solvent and the size and charge of the ions play large roles: *the lower the dielectric constant, the smaller the ions (including tightly-bound solvent molecules), and the larger the charges, the more a salt is associated into ion pairs*. There can be additional effects due to mutual polarization of the ions at short range, or even covalent bond formation. Bjerrum treats the solvent as a continuous dielectric, and derives a critical distance, $q = z_i z_j e^2/2\epsilon_r kT$, at the minimum of the curve of probability for finding an ion of opposite charge as a function of distance from the central ion, seen in Fig. 2.9. If the distance separating two ions i, j is less than q the pair is 'bound'; if greater it is free. If the distance of closest approach, a, exceeds q, pairing does not occur. In Fig. 2.10 an arbitrarily-chosen function of the constant for the formation of ion pairs in solutions of tetrabutylammonium iodide, K, is plotted versus the Clausius–Mosotti function of the dielectric constant. As the dielectric constant of the solvent increases, the constant decreases, becoming less than unity at a value of ϵ_r of 40$\pm$ about 3 units, making $a = 700 \pm 50$ pm, not far from the sum of the radii of the bare tetrabutylammonium and iodide ions.

Fuoss and Krauss treat an ion pair as existing if the ions, including any solvent molecules firmly bound to the ions, are in contact, and not existing if one or more additional solvent molecules intervene. The probability of the ions being separated by a fraction of a solvent molecular diameter is low. Their resulting calculated dissociation constants are given by eqn 2.25.

$$K = \frac{\alpha^2 y_\pm{}^2 c}{1-\alpha} = \frac{3000}{4\pi N a^3 \exp b}; \quad b = \frac{|Z_1 Z_2| e^2}{\epsilon_r kTa} \tag{2.25}$$

They are not so very different from Bjerrum's, which is a comfort, for explicitly accounting for solvent molecules is often difficult mathematically, and we should like to avoid it when we may. The difference is hard to test

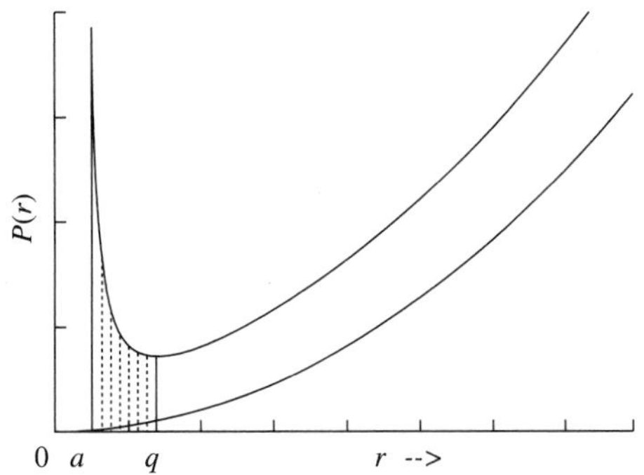

Fig. 2.9 Bjerrum probability distribution. Probability that a species of given charge (0 or −1) will be found at a distance r from a central ion of charge +1. A negative ion inside the distance q is considered paired with the central cation. If the distance of closest approach is a, the shaded area under the curve, $a < r < q$, is proportional to the probability that a pair will exist. If $a > q$, pairing does not occur. After Bjerrum (1926).

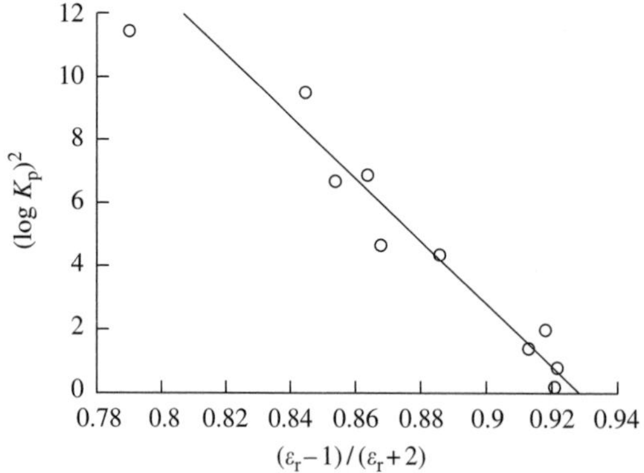

Fig. 2.10 Ion-pairing constant for tetrabutylammonium iodide in various solvents (Mayer *et al.* 1975). Plot of $(\log_{10}K)^2$ versus a function of dielectric constant.

experimentally, but some of Fuoss and Krauss's data (1933) on the salt tetra-isoamylammonium nitrate suggest that their equation is slightly better than Bjerrum's.

Winstein *et al.* (1954) distinguish contact ion pairs, solvent-separated ion pairs, and widely-separated ion pairs, that is, free ions. The distinction between contact and solvent-separated ion pairs was demonstrated experimentally by Smithson and Williams (1958), Hogen-Esch and Smid (1966), and Buncel and Menon (Buncel and Menon 1979; Buncel *et al.* 1979). Figure 2.11(a) shows the visible absorption spectra of (triphenylmethyl)lithium (**3**) in diethyl ether, in tetrahydrofuran, and in 1,2-dimethoxyethane at room temperature. The large shift of the absorption maximum was interpreted as indicating that the solute is present predominantly as contact ion pairs in ether, but solvent separated in THF and in dimethoxyethane. Changes were observed in the absorption spectrum in ether as the temperature was lowered from −9.5 to −50 °C (see Fig. 2.11(b)). The absorption maximum near 450 nm appears to be due to the contact ion pair, and that near 500 nm to the solvent-separated pair. As the temperature is lowered, solvent-separated ion pairs are favoured. Buncel and Menon calculate for K, the ratio of the concentrations of solvent-separated to contact pairs, the value 0.1 at 25 °C, and estimate $\Delta H \approx -12\ \mathrm{kJ\ mol^{-1}}$ and $\Delta S \approx 46\ \mathrm{J\ mol^{-1}\ K^{-1}}$ (each ± 10–20%).

Clearly the notion of ion pairs is well established (Szwarc 1968, 1972; Jones 1973). In any but very polar solvents, it should therefore be assumed that ions invoked in mechanisms may exist partly or chiefly as pairs, which will have an effect on the form of the rate law for a reaction (see Problem 2.2). Szwarc (1972: p. *v*) emphasizes the importance of ion pairing in mechanisms. The presence or absence of ion pairing may lead to very different rates, or even different courses of reaction. While the reactivity of free ions is generally greater than that of ion pairs, in a study by Dunn and Buncel (1989) of the ethanolysis of p-nitrophenyldiphenyl phosphinate (**4**), the kinetic behaviour of different alkai-metal ethoxides indicated that $\mathrm{EtO^-M^+}$ *ion pairs are more reactive than the free* $\mathrm{EtO^-}$ *ions.*

In more concentrated solutions, further aggregation to *ion triplets, quadruplets*, etc., may occur (e.g. see, Erdey-Grúz 1974: 434–436.) This and the pairing of ions that have unequal charges make for mathematical

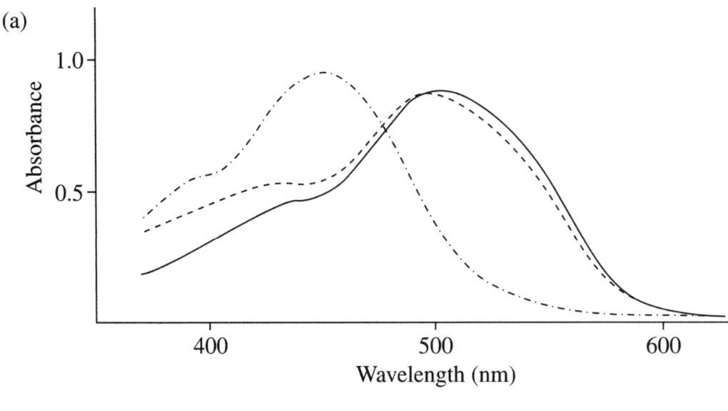

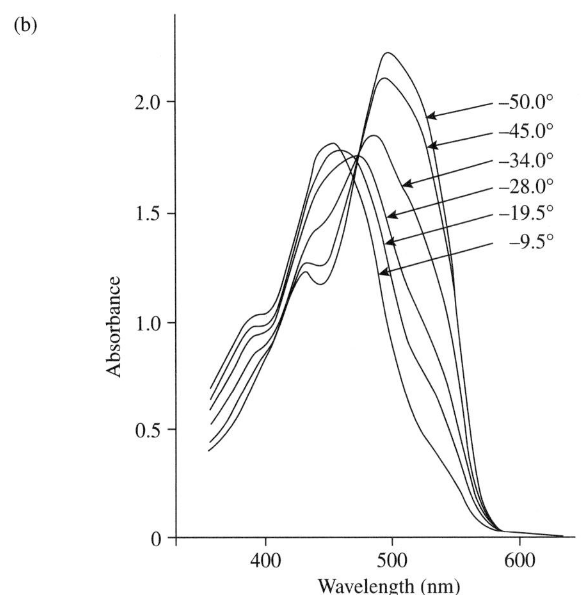

Fig. 2.11 (a) Visible absorption spectra of (triphenylmethyl)lithium in ether (dash-dot), THF (solid) and 1,2-dimethoxyethane (dashes). Reproduced from Buncel and Menon (1979) with permission. (b) Temperature dependence of visible absorption spectrum of (triphenylmethyl)lithium in ether. Reproduced with permission from Buncel and Menon (1979).

complications, especially in the interpretation of the dependence of the conductance on concentration, and in kinetics. Balakrishnan *et al.* (2001) found that in degradation of the organophosphorus pesticide, fenitrothion (**5**), in ethanol in the presence of alkali-metal ethanolate, one pathway was catalysed by free ethanolate ion, but also by ion pairs and by $[2M^+ \cdot 2EtO^-]$ quadruplets, the last being the most effective.

5

In what may be seen as an extreme example of ionic association occurring in concentrated solution, during a study of the behaviour of salt solutions in electrospray ionization mass spectrometry (Hao *et al.* 2001) two distinct kinds of ionic species were detected, of mass/charge ratio corresponding to $(H_2O)_nNa^+$ and $(NaCl)_mNa^+$ (among others). In the early stages of evaporation of a charged droplet of a dilute aqueous NaCl solution *in vacuo* species of the first kind appear to be formed (Thomson and Iribarne 1979); in later stages, the predominant species are of the second kind (Röllgen *et al.* 1984). In the hydrated species, the abundance of successive species falls away in a fairly regular manner as the number of water molecules present increases. The salt clusters, however, favour certain values of m ('magic numbers') that correspond to clusters of high symmetry (Doye and Wales 1999), notably $m = 13$, which corresponds to a $3 \times 3 \times 3$ cubic cluster, containing 13 Cl^- and 14 Na^+ ions.

When the same experiment was tried with dilute solutions of sodium carbonate in methanol (Hao and March 2001), a surprising result was the appearance of species $(NaOCH_3)_nNa^+$. In studies on buffer solutions based on phosphate with initial pH values on either side of neutrality, it had been observed that the pH appeared always to shift away from 7 as the droplet evaporated, judged by the species detected in the mass spectrum. The appearance of the species containing $NaOCH_3$ was nevertheless unexpected.

Davies (1962: ch. 9) gives a large table of the dissociation constants for a number of salts in a variety of solvents. He points out that the solvents appear to fall into two classes, which he calls *levelling* and *differentiating* (these terms are also used in describing the effects of solvents on acids and bases; see Chapter 3). The levellers are principally the hydroxylic solvents. In these solvents inorganic salts tend to be strong electrolytes, and where a comparison of the alkali metal salts of one acid is possible, the Li salt tends to be strongest and the Cs salt the weakest. The data for nitrates in methanol, $\epsilon_r = 32.6$, are plotted against the reciprocal of the crystal radius of the cation in Fig. 2.12. The probable reason for this order is that the smaller cations are able to bind solvent molecules more tightly, making a larger primary solvation sphere, so that $Li(solv)^+$ is effectively the largest and $Cs(solv)^+$ the smallest (Pregel *et al.* 1995). All are nevertheless roughly comparable in size, except often lithium, which, being unique in having only a helium-type core, is able to bind four (usually) molecules exceptionally tightly.

In the differentiating solvents, solvation is much less important, and the intrinsic differences in ionic size are not masked. The dissociation constants tend to be smaller than in levelling solvents, and the order is reversed: for example, the alkali metal picrates (salts of 2,4,6-trinitrophenol) in acetonitrile ($\epsilon = 36$), also plotted in Fig. 2.12.

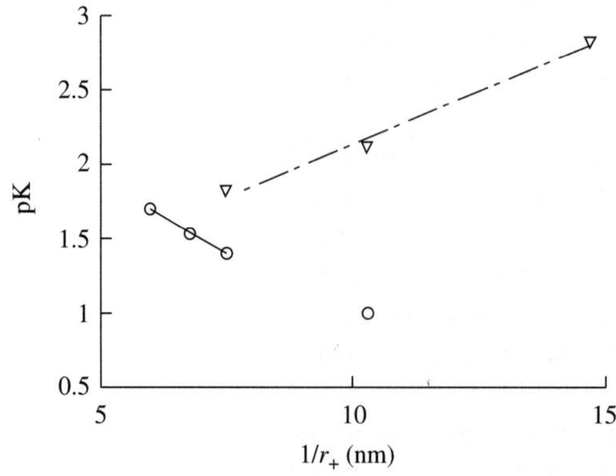

Fig. 2.12 Ionic radius and ion-pair dissociation: pK_{diss} for K, Na, Li nitrates in methanol (circles) and Cs, Rb, K, Na picrates in acetonitrile (triangles), versus reciprocal of cation crystal radius. Data from Davies 1962: 95, 96. The value of *pK* for $NaNO_3$ in methanol is reported as <1.

A salt having two large ions of low charge, such as tetraethylammonium picrate (**6**), should show nearly perfectly the electrostatic effect, uncomplicated by the molecular nature of the solvent. Davies (1962: 97) shows data, claimed to be very precise, for the pK of this salt in nitrobenzene, acetone and 1,2-dichloroethane, plotted versus $1/\epsilon$, giving a nearly perfect straight line, as required by the theory of Fuoss and Krauss, eqn 2.23. The tetraethylammonium halides and the alkali picrates in the same solvents give poorer graphs of this kind, because of the differing degrees of primary solvation of the small ions.

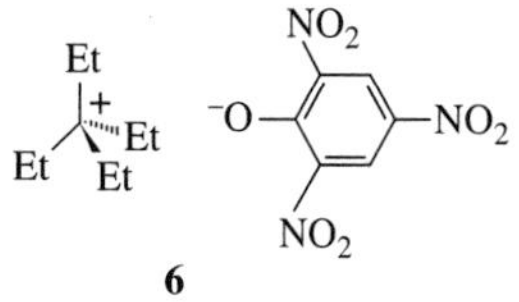

Et Et Et Et NO_2 ^{-}O NO_2 NO_2

6

2.10 Mixed solvents

Mixed solvents present both an opportunity and a problem. On the one hand, it is possible, by varying the proportions of components of a binary mixture, to vary certain properties continuously over a wide range. For example, water and 1,4-dioxane are miscible in all proportions. Mixtures can be prepared with any value of the dielectric constant from 2.2 to 78.5 at 25 °C. Mixtures of acetonitrile ($\epsilon_r = 36.2$) with dimethylformamide (36.7) are nearly *isodielectric*; other properties can be varied while keeping the dielectric constant nearly unchanged. The liquid range of a solvent can be extended slightly upward by admixture of a cosolvent with which the first forms a high-boiling azeotrope, for example, chloroform (b.p. 61.0 °C) and acetone (56.3 °C) form an azeotrope boiling at 63.4 °C. Since adding an impurity to a liquid usually lowers its freezing point, the liquid range is extended downward, the eutectic temperature being below the freezing point of either component. For example, the eutectic mixture of LiCl (m.p. 613 °C) and KCl (m.p. 776 °C) melts at 355 °C. Binary salt mixtures that are liquid at room temperature are available (Barthel and Gores 1994).

On the other hand, it is not always clear what the environment of a solute species in a mixed solvent may be. Preferential solvation of ions by one component is common. Viscosity measurements of dilute salt solutions in water–methanol have been interpreted (Stairs 1976, 1979) as showing free energies of preferential binding to various alkali halides of about 1 kJ mol^{-1} for water over methanol, without distinguishing between the cation and anion, and without detecting any trend. Strehlow and Schneider (1969) have distinguished cases in which the cation and anion are both preferentially solvated by the same component of a solvent mixture (*homoselective*) or by different components (*heteroselective*). Heteroselectivity can have interesting effects, such as enhancement of solubility of a salt in mixed solvent over that in either component, for instance, silver sulphate in water–acetonitrile. These solvents show partial miscibility below about −1 °C. Addition of 0.1 mol L^{-1} of the homoselectively solvated salt $NaNO_3$ raises the critical solution temperature (defined for a binary system showing partial miscibility as the temperature above which the components are miscible in all proportions) nearly 3°, while the heteroselectively solvated salt $AgNO_3$ lowers it more than 6° (Strehlow and Schneider 1971). Phase separation is favoured by homoselective solvation; it is opposed by heteroselective solvation. Ràfols *et al.* (1997) have examined the effects of preferential solvation in a number of systems on $E_T(30)$, π^*, α, and β.

Covington and Newman (1976) report the results of their NMR study of solutions of alkali halides in binary mixtures of water with seven other solvents, and UV-VIS measurements on alkali-metal bromides, iodides and nitrates in water–acetonitrile, water–DMSO and acetonitrile–methanol mixtures. Some interesting trends appear. For instance, in the nearly isodielectric binary solvent system water/hydrogen peroxide, the preferential solvation of the alkali metal ions by water over hydrogen peroxide decreases as the ionic radius increases; for the halide ions (as far as can be seen, since Br^- and I^- cannot be studied in the presence of peroxide) the trend is in the opposite direction. For water–methanol solutions, they find free energy differences favouring water at sites on Na^+, Rb^+, Cs^+, and Cl^- of 1.28, 0.89, 0.66, and 0.96 kJ mol^{-1} and for F^- of -0.15 kJ mol^{-1} (i.e. favouring methanol).

A measure of selective solvation of single ions is the *isosolvation point*, that is, the solvent composition at which the mean numbers of molecules of the two solvent components bound to an ion are the same. Rahimi and Popov (1976) found by $^{109}Ag^+$ NMR that the isosolvation point for Ag^+ in water-acetonitrile was at 0.075 mole fraction of acetonitrile. This means that solvation by this component is preferred by about 6.3 kJ mol^{-1} over water.

A complication in preferential solvation exists in the possibility of discriminatory effects (discussed further in Chapter 3) or their contrary. In the simple case represented by the equations:

$$MA_2 + B \rightleftharpoons MAB + A$$
$$MAB + B \rightleftharpoons MB_2 + A$$

if the composition of the solvent is at the isosolvation point, that is, the concentrations of MA_2 and MB_2 are equal, the fraction of M present as MAB at equilibrium should be $\frac{1}{2}$. If, however, discrimination exists (in the sense, 'You can't join my club unless you are like me') this fraction will be smaller. *Antidiscrimination* ('affirmative action'?) will make it larger.

Nevertheless, a number of authors report studies of solvent effects in mixed solvent systems. Amis and coworkers (Amis and LaMer 1939; Amis and Price 1943), guided by Scatchard's (1932) theory of the effect of dielectric constant on rates of reactions between ions, constructed plots of the logarithm of the rate constant versus $1/\epsilon_r$ for the reaction of tetrabromophenol–sulphonephthalein (**7**) with hydroxide in water–methanol and water–ethanol, and for the reaction of ammonium ion with cyanate in water–methanol and in water–ethylene glycol. In each case they found that the curves were essentially linear in the water-rich region, down to dielectric constants of 55–45. As the amount of the organic component of the solvent increased further, curvature appeared, attributed to changes either in the mechanism or in the state of primary solvation of the reacting ions.

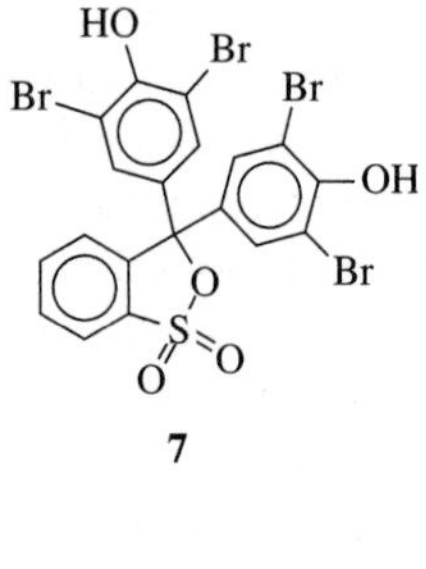

7

In highly structured solvents, such as water and alcohols, structural changes with composition may be manifested in physical properties. It has been suggested (Werblan *et al.* 1971) that small additions of hydroxylic solvents to water enhance the special structure of water, up to a point. The *B* coefficient of viscosity for solutions of alkali halides in water/methanol shows a minimum at 0.15–0.2 mole fraction of methanol (Stairs 1979), presumably because the structure-breaking effect of the ions finds the most structure to break in this range of solvent composition. The activation parameters $\Delta_{\ddagger}H$ and $\Delta_{\ddagger}S$ for the reaction of Ni^{2+} with 2,2′-bipyridyl in water/ethanol have strong maxima at 0.08 mole fraction of ethanol (Tobe and Burgess 1999). The effect largely cancels in $\Delta_{\ddagger}G$, as is often observed, but their plot of $\Delta_{\ddagger}H$ versus $\Delta_{\ddagger}S$ for the similar system with methanol as cosolvent shows curious kinks.

Solvolysis studies of alkyl halides in mixed solvents are common (see, e.g., Buncel and Millington 1965; Blandamer *et al.* 1985). Grunwald and Winstein (1948) introduced the parameter Y, a measure of *ionizing power* based on the rate of solvolysis of 2-chloro-2-methylpropane (*t*-butyl chloride). It is a measure of the ability of the solvent to stabilize the ions that are generated in the unimolecular, rate-determining first step:

$$(CH_3)_3CCl \rightarrow (CH_3)_3C^+ + Cl^- \quad k_1 \text{ (slow)}$$
$$(CH_3)_3C^+ + 2HX \rightarrow (CH_3)_3CX + H_2X^+ \quad k_2 \text{ (fast)}$$

HX represents an amphiprotic solvent. Since the first step is slow, $k = k_1$. Y is defined by eqn 2.26:

$$Y = \log\left(\frac{k_S}{k_0}\right) \qquad (2.26)$$

where k_S is the rate constant for solvolysis in solvent **S**, and k_0 is the constant for the reference solvent, which is ethanol/water 80 : 20 v/v. Fainberg and Winstein (1956) report values of Y for a number of aqueous–organic and a few organic–organic mixed solvent systems. In water–ethanol mixtures, Y is not a linear function of the recorded volume percent, but is nearly linear on a mole percent scale.

The solvatochromic parameter $E_T(30)$, introduced by Reichardt and coworkers (Dimroth *et al.* 1963) is the energy of a transition observed in the UV-VIS absorption spectrum of the substituted pyridinium-*N*-phenoxide dye (**8**), 30th of a number tried. It has been measured in a great many single and mixed solvents. Reichardt (1994) provides a list of references. He points out that the dependence of this and other solvatochromic shifts on composition of binary solvent mixtures is not simple, being linear neither in mole fraction nor in volume fraction.

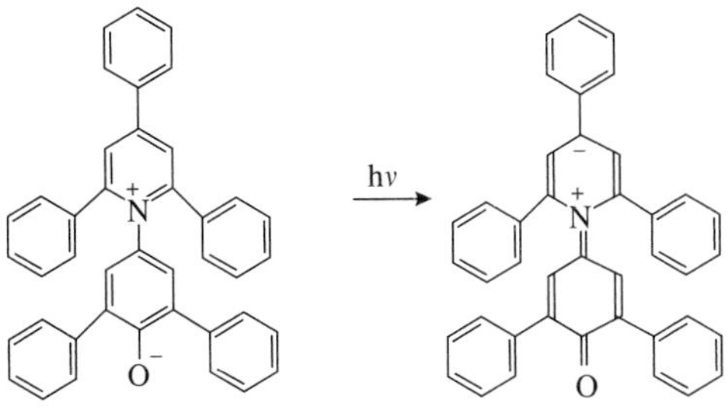

8

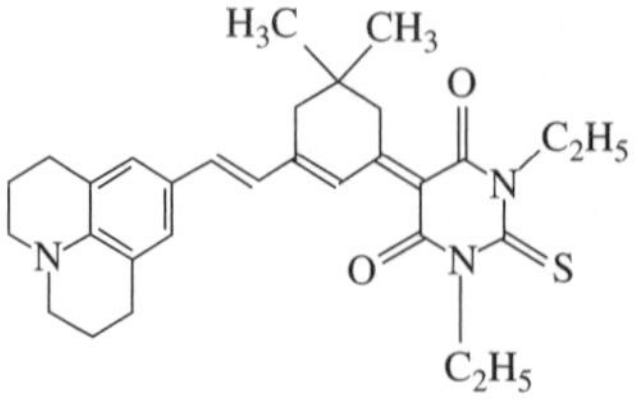

Brooker *et al.* (1965) report measurements of χ_R and χ_B (solvent polarity measures based on $\pi \rightarrow \pi^*$ transitions of two meropolymethine dyes, **9** and **10**) for a number of mixtures among water, methanol, 2,6-lutidine, 1,4-dioxane and isooctane, which on volume fraction scales required cubic polynomials to be approximately fitted (by the present authors). In some instances, parameters such as $E_T(30)$ pass through a maximum. This can occur if the two components can, through hydrogen bonding, form a complex that is more polar than either alone, such as in DMSO–alcohol mixtures (Maksimović *et al.* 1974). These effects, arising as they do from preferential solvation of the solvatochromic dyes, may not reflect the effects of varying solvent composition on reactions, unless the preferential solvation of reactants and products or activated complexes parallels the preferential solvation of the ground and excited states of the indicator dyes. It is apparent that conclusions drawn from effects on rates or equilibrium of solvent composition in mixed solvents cannot be confidently attributed, for instance, to changes in the bulk dielectric constant, or to any other single property of the mixed solvent.

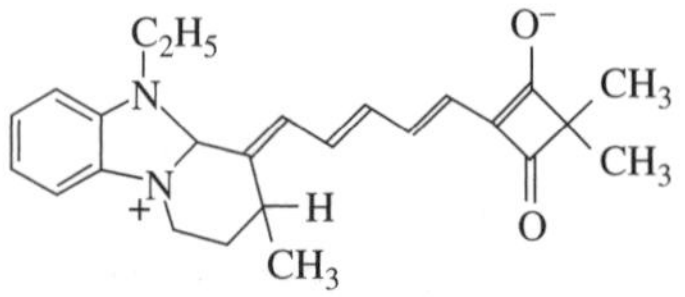

9

10

Langhals (1982*a,b*) fitted values of $E_T(30)$ for a number of binary mixtures by eqn 2.27, using a non-linear least-squares procedure.

$$E_T(30) = E_D \ln\left(\frac{c_p}{c*} + 1\right) + E_T^0(30) \qquad (2.27)$$

Here $E_T(30)$ represents the polarity of the binary mixture, and $E_T^0(30)$ that of the pure, less polar component; c_p is the molar concentration of the more polar component, and c^* and E_D are adjustable parameters. Langhals showed that an equation of this form can also express the concentration dependence of several other measures of solvent polarity in binary mixtures. This equation predicts that, above a certain concentration, $E_T(30)$ should be linear in ln c_p, as observed. Langhals also suggested (1982*b*) that it can be used in homologous series, by considering a linear alcohol $H(CH_2)_nOH$, for instance, as a mixture of one mole of the polar H–OH group with n moles of the non-polar CH_2. This means that c_p is equal to the reciprocal of the molar volume of the alcohol, that is, $c_p = \rho/M$. Guided by this concept, he obtained a plot of $E_T(30)$ linear in ln c_p, or rather one with two distinct linear sections.

The present authors note that this same concept, translated into mole fraction terms, gives for the mole fraction of the polar group, $x_p = 1/(1+n)$, and that of the non-polar chain segments, $x_0 = n/(1+n)$. If the effect is assumed to be additive, eqn 2.28 results:

$$E_T(30) = \frac{E_p}{1+n} + \frac{nE_0}{1+n} \qquad (2.28)$$

Here E_p and E_0 should represent the values of $E_T(30)$ appropriate to the polar group and the CH_2 group, respectively. Figure 2.13 is Reichardt's (1994) figure 1, with the addition of the curve calculated from eqn 2.28, with $E_p = 63.37$ and $E_0 = 46.27$. These values may be compared with the values for water (63.1) and for a normal alkane (31.0–31.1). A similar fit of data for nitriles (three points only) gave $E_p = 51.8$ and $E_0 = 39.4$. This latter value of

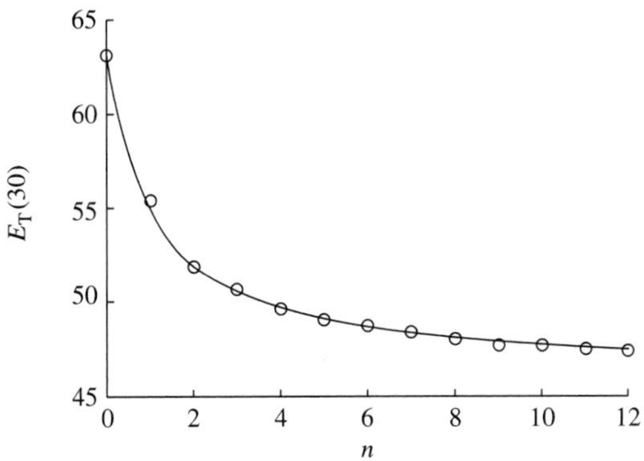

Fig. 2.13 $E_T(30)$ values for water and normal alcohols $H(CH_2)_nOH$. The curve is represented by: $E_T(30) = (E_p + nE_0)/(1+n)$, with $E_p = 63.37$ (the polar group contribution), $E_0 = 46.27$ (the methylene contribution) Redrawn from Reichardt (1994), with permission.

E_p may be taken as an estimate of the value of $E_T(30)$ for HCN, not otherwise known. Why the E_0 values are consistently higher than the alkane values is not clear.

For graphical determination of the constants E_p and E_0, eqn 2.28 may be linearized in the form:

$$(1+n)E_T(30) = E_p + E_0 n \tag{2.29}$$

A plot of $(1+n)$ $E_T(30)$ versus n for these straight-chain alcohols was linear from $n = 0$–12 , with a correlation coefficient $r = 0.999$. Secondary, tertiary and branched-chain alcohols all fell below the line, showing that, in general, correlations between properties are greatly dependent on structure.

2.11 Salt effects

A reaction medium consisting of a pure solvent plus added salt is a mixed solvent of a special kind. Reactions involving no ions or only one, among reactants or products, or in the transition state, should not be subject to large salt effects. Where both reactants in a bimolecular process are charged, however, electrostatic contributions to the energy of activation should exist. These contributions are, according to the Debye–Hückel theory, reduced by the presence of other ions.

The magnitude of the kinetic salt effect for a bimolecular reaction depends strongly on the charges on the reacting species. Equation 2.30, derived from the Debye–Hückel theory, has been shown to represent most experimental data well (Atkins 1998: 836).

$$\log(k/k_0) = 2Az_+z_-I^{1/2} \tag{2.30}$$

Here A is the constant in the Debye–Hückel limiting law. For water at 25 °C, $A = 0.51$. Following Pethybridge and Prue (1972), a plot of the logarithm of the rate constant against $I^{1/2}/(1+I^{1/2})$, a better abscissa than simply $I^{1/2}$ (cf. the Davies approximation, eqn 2.22), should yield a straight line of slope equal to $2A$ times the product of the charges of the reacting ions, z_+z_-. Tobe

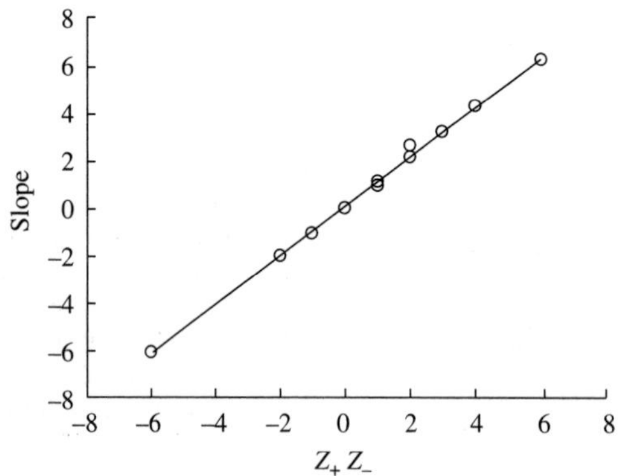

Fig. 2.14 Salt effect: slope of graph of $\log(k/k_0)$ versus $I^{1/2}/(1+I^{1/2})$, where I is ionic strength, for eleven reactions between ions of different charge type in aqueous solution at 25 °C, versus the product of the charges on the reacting ions. Data from Tobe and Burgess (1999: 359).

and Burgess (1999: 359) show such plots for a variety of reactions of mixed character, organic and inorganic. Figure 2.14 shows a plot of the slopes of the lines in their figure versus the charge product, showing the trend clearly, with the correct slope. Tobe and Burgess caution, however, that specific salt effects can occur that may obscure the electrostatic effect, especially where organic ions are present.

Problems

2.1. Using any form of the Debye–Hückel Equation, estimate how dilute a solution of sodium chloride must be for the solubility of barium sulphate in this solution at 25 °C to be within 5% of the value calculated from the solubility-product constant.

2.2. Consider a reaction between ions A^+ and B^- that proceeds through the ion pair A^+B^-, that is,

$$A^+ + B^- \rightleftharpoons A^+B^-; \text{rapid equilibrium}, K = \frac{[A^+B^-]}{[A^+][B^-]}$$

$$A^+B^- \rightarrow \text{product; slow}, k.$$

If C is the analytical concentration of the salt AB, show that

(i) the general expression of the rate law is

$$\text{rate} = \frac{d[P]}{dt} = \left(\frac{k}{2K}\right)\left((1+2KC) - (1+4KC)^{1/2}\right)$$

(ii) the order of the reaction with respect to C is

$$\text{ord} = \frac{d\ln(\text{rate})}{d\ln C} = \frac{2KC(1-(1+4KC)^{-1/2})}{(1+2KC)-(1+4KC)^{1/2}}$$

(iii) Find the limiting values of the order at small and large values of KC.
(iv) At what value of KC is the order $\frac{3}{2}$?

2.3. Show by a statistical argument that in a system containing the species MA_2, MAB, and MB_2 at equilibrium, with equal concentrations of MA_2 and MB_2 (at the isosolvation point), the fraction of M present as MAB should be $\frac{1}{2}$ in the absence of discrimination. Assuming this is true for the silver ion in water/acetonitrile and that the coordination number of Ag^+ is 2, calculate the equilibrium constant and the value of ΔG^0 at 298 K for each of the following reactions, using the finding of Rahimi and Popov (1976) that the isosolvation point is at 0.075 mole fraction of acetonitrile.

$$Ag(H_2O)_2^+ + CH_3CN \rightleftharpoons Ag(H_2O)(CH_3CN)^+ + H_2O$$

$$Ag(H_2O)(CH_3CN)^+ + CH_3CN \rightleftharpoons Ag(CH_3CN)_2^+ + H_2O$$

3 The solvent as participant

3.1 Specific interactions

Negative deviations from Raoult's Law imply that the interaction energy of two unlike molecules is greater than the mean of the two 'like–like' interactions (we have seen that mere difference between the molecules leads to the opposite effect). Specific interactions that could cause negative deviations include hydrogen bonding, Lewis acid–base reaction, and charge transfer.

Solvation and *solvolysis* must be distinguished. The former term implies that the solute remains intact, though with one or more attached solvent molecules. The strength of attachment may vary widely, and may involve only van der Waals interactions as described in Section 2.1, but may include specific interactions: through hydrogen bonding (Section 3.2), through the formation of covalent bonds by the donation and acceptance of pairs of electrons, usually discussed under the heading of Lewis acid–base chemistry (Section 3.6), or through charge transfer. This last we will not discuss further than to note the familiar example of molecular iodine, which in the vapour and in solution in solvents such as tetrachloromethane is violet in colour, but in benzene, for instance, is brown, owing to the ability of benzene to receive electron density, transferred from a non-bonding orbital of I_2 into its anti-bonding π^* molecular orbitals.

Solvolysis, on the other hand, describes a reaction between the solute and the solvent, in which the identity of the solute is lost. In sulphuric acid, typical of strongly acidic solvents, most salts, if they are at all soluble, react to produce the conjugate acid of the anion and the sulphate of the cation (which may precipitate), for example:

$$2Fe(NO_3)_3 + 3H_2SO_4 \rightarrow Fe_2(SO_4)_3(s) + 6HNO_3$$

In aqueous solution, many salts are to different extents solvolysed (hydrolysed), the solution becoming acidic if the cation is small or highly charged:

$$Mg^{2+} + 7H_2O \rightarrow [Mg(H_2O)_6]^{2+} + H_2O \rightleftharpoons [Mg(H_2O)_5OH]^+ + H_3O^+$$

or basic if the anion is derived from a weak acid:

$$CH_3COO^- + H_2O \rightleftharpoons CH_3COOH + OH^-$$

An example of solvolysis important in physical organic chemistry is the solvolysis of *t*-butyl chloride (2-chloro-2-methylpropane), the rate of which is the basis of the Grunwald and Winstein (1948) parameter *Y*, described in Section 2.9.

3.2 Hydrogen bonding

Species in which a hydrogen atom is attached to an atom that is intrinsically electronegative, such as oxygen, or that is made electronegative by other attached atoms, as is the carbon atom in chloroform, possess a local (bond) electrical dipole, with hydrogen bearing a partial positive charge, owing to withdrawal of electron density toward the electronegative atom. This also results in diminution of the van der Waals radius of the hydrogen, so it is able to approach a centre of negative charge on another molecule. The electrostatic attraction is accompanied by some orbital involvement. The resulting attractive interaction is termed a *hydrogen bond*, and the molecule bearing this hydrogen is termed a *hydrogen bond donor* (HBD). A *hydrogen bond acceptor* (HBA) is a molecule that possesses a site where an unshared pair of electrons on an atom bearing a partial negative charge is sterically accessible to the approaching HBD molecule. Intramolecular hydrogen bonds (Ahlberg *et al.* 1997) are also possible. Typical hydrogen bonds, for instance those between water molecules and between the OH groups in alcohols, which have both HBA and HBD character, have strengths in the region 15–20 kJ mol^{-1}. Hydrogen bonds in amines are weaker, while those involving fluorine (for instance, in the ion FHF^-) are stronger.

The structure and solvent properties of water are dominated by hydrogen bonding. The high melting and freezing points of water relative to other molecules of comparable size, such as methane, are obvious consequences of the ability of water to form a network in three dimensions. In ice the network is essentially complete; in the liquid it is imperfect, but persists to temperatures above the melting point, and is responsible for the decrease in density of water as the freezing point is approached. A large part of water's solvent power for salts derives from its ability to form hydrogen bonds with anions. Its miscibility with 1,4-dioxane and with acetone result from hydrogen bonding with the oxygens in these HBA molecules.

Hydrogen atoms in hydrocarbons are essentially unable to take part in hydrogen bonding, for the electronegativities of carbon (2.5) and hydrogen (2.1) are not sufficiently different, but in compounds in which a carbon bears electronegative substituents, or is conjugatively influenced by electron-withdrawing groups, the situation may change. Chloroform is a weak HBD substance and acetone is a moderately strong HBA. Binary mixtures of these exhibit negative deviations from Raoult's law, sufficient to cause a maximum in the boiling point (azeotrope) at about 0.8 mole fraction of chloroform.

Hydrogen bonding may be viewed as a sort of tentative acid–base interaction. If a HBD molecule made an outright gift of a proton, instead of just sharing it, we should call it an acid. The receiving molecule would then be acting as a base. In the next sections we take up the consequences of this, and consider the Brønsted–Lowry and Lewis views of acids and bases.

3.3 Acids and bases in solvents

In any discussion of acids and bases in solvents it is necessary to distinguish between solvents that are themselves capable of acting as acids or bases and those that cannot do so. In the context of the Brønsted–Lowry definition, molecules that contain no hydrogen clearly cannot act as acids, but it is difficult to conceive a molecule that could in no circumstances be protonated. For example, the species CH_5^+ is well known from mass-spectrometric studies, and protonation of alkanes has been demonstrated in superacid solutions (Fabre *et al.* 1977; Olah *et al.* 1985). Similarly in the Lewis context, it is difficult to conceive a solvent that could in no circumstances become involved in electron-pair sharing either as donor or as recipient. Clearly, though, solvents that are too weakly acidic or basic to interact appreciably with any given set of Brønsted or Lewis acids and bases *are* conceivable. They are best called 'inactive' or 'inert', bearing in mind that these are relative terms. Aprotic or inert solvents do not require us to add anything to the considerations of the previous chapter, as they function as mere media. *Active solvents*, on the other hand (the term may as well include proton or electron-pair donor or acceptor solvents) are able to participate in reactions, either as reactants or products, or, more subtly, as catalysts. We shall consider proton donor and acceptor solvents first.

3.4 Brønsted–Lowry acids and bases

In this scheme, an acid is any species (not 'substance') that can donate a proton to another species, and a base is a species that can receive a proton. (see, e.g., Stewart 1985.) Since bare protons cannot exist for any appreciable time in a liquid, acids and bases always react in mutual fashion by proton transfer:

$$HA + B \rightleftharpoons A + HB$$

Charges have been omitted from this equation; A and B will always be one unit more negatively charged than HA and HB. HA and A are termed *conjugate* acid and base. Either HA or B may represent a solvent molecule. *Strength* of an acid is relative; if HA is a stronger acid than HB, the equilibrium will lie to the right. This is exactly equivalent to saying that B is a stronger base than A.

Solvents that are *amphiprotic*, that is those that can both give and receive a proton, are capable of *autoprotolysis* or *autoionization*, for example, for water:

$$2H_2O \rightleftharpoons H_3O^+ + OH^-$$

or more generally, if we represent the solvent molecule by HS:

$$2HS \rightleftharpoons H_2S^+ + S^-$$

In the pure solvent, the two ions will be present in equal concentrations, but the presence of a foreign acid or base is signalled by the presence in solution of an excess of one or the other of the characteristic ions of the solvent. This is the basis of the Arrhenius definition of an acid as a substance (not 'species')

Table 3.1 Autoionization Constants of selected solvents[1]

Solvent	t/ (°C)	pK_{ip}	pH_n
Water	25	14.0	7.0
Ammonia	−33	27	13.5
1,2-diaminoethane	25	15.3	7.7
Methanol	25	16.7	8.4
Ethanol	25	18.9	9.5
Hydrogen fluoride	0	13.7	6.9
Sulphuric acid[2]	25	3.57	1.8
Phosphoric acid	25	0.9	0.5
Acetic acid	25	14.5	7.2
Formic acid	25	6.2	3.1
Fluorosulphonic acid	25	7.4	3.7

[1] Data largely from Jolly (1970; p 99); see also Fabre *et al.* (1982).
[2] The autoionization of sulphuric acid is complicated by the existence of two modes: ionic dehydration: $2H_2SO_4 = H_3O^+ + HS_2O_7^-$; $K_{id} = 5.1 \times 10^{-5}\,mol^2\,kg^{-2}$ autoprotolysis: $2H_2SO_4 = H_3SO_4^+ + HSO_4^-$; $K_{ap} = 2.7 \times 10^{-4}\,mol^2\,kg^{-2}$ (Gillespie and Robinson 1965; Cerfontain 1968).

which when dissolved in water yields ‘hydrogen ions’ (hydrated H_3O^+ ions). An acid substance is ‘strong’ in a given solvent if it can donate protons to the solvent to such an extent that a c molar solution of the substance in the solvent yields a concentration of the *solvo-acid* ion, $[H_2S^+] = c\,mol\,L^{-1}$, approximately (here we have used the convention that the square brackets [ ] represent the actual concentration of a species, while c represents the formal concentration of a substance). If $[H_2S^+] \ll c$, the acid is ‘weak’ in that solvent. Since $[H_2S^+]$ can never exceed c, the strongest acids all appear equally strong. This is called *levelling*[1]. *Differentiating* solvents, on the other hand, are those that are weak enough bases so that the reaction of most acids with the solvent is incomplete, and different acids react to differing degrees. In water, hydrogen chloride and acetic acid are differentiated, but in liquid ammonia both are strong.

The activities (or, approximately, concentrations) of the solvo-acid and *solvo-base* (solvent minus a proton) are linked through the autoprotolysis or *ion-product constant*:

$$K_{ip} = [H_2S^+][S^-]$$

which for water at about 25 °C is $1.00 \times 10^{-14}\ mol^2\,L^{-2}$. Some values for other solvents are listed in Table 3.1. There is no direct relation between the

[1] Levelling up, that is, like the Portuguese revolution of 1910, which made everyone ‘Your Excellency’, in contrast to the French one, which levelled everyone down to ‘Citizen’ (Maurois 1927).

autoprotolysis constant and the acid strength; rather it depends on the difference between the acidities of the neutral molecule and the monoprotonated ion. Since the concentrations of both ions must be equal in the pure solvent, each must be equal to $K_{ip}^{1/2}$, and the 'neutral point' on a pH scale for any solvent is $pH_n = -\frac{1}{2}\log_{10}K_{ip}$: 7.00 for water at 25 °C, about 15 for ammonia at −33 °C, about 7.2 for acetic acid at 25 °C. If the pH is greater than pH_n, protons have been transferred from the solvent to the solute; the solution is basic. If the pH is less than pH_n, protons have been transferred to the solvent; the solution is acidic. The pH scales are thus different in each solvent. So are the related 'relative acidity index' scales, $A = pH_n - pH$, which place the neutral point for each solvent at zero, and do away with the confusing contrary motion of acidity and pH (Stairs 1978, 1983).

3.5 Acidity functions

Following the pioneering work of Hammett (1932, 1970), in establishing the acidity functions, H_0, H_+, H_-, etc., it became apparent that organic bases of different structure lead to different scales. Efforts at reconciling them date back at least to Bunnett and Olsen (1966). Cox and Yates (1983) list 39 different symbols, representing 425 more-or-less distinct scales, depending on different classes of indicator bases and suitable for use in different solvent systems and acidity/basicity ranges.

It would, of course, be desirable to define a single scale that measures proton activity, a_{H^+}, such that in dilute aqueous solution it reduces to the accepted scale: $pH = -\log_{10}(a_{H^+})$. Writing for a particular base A in acid solution the equilibrium:

$$AH^+(solv) \rightleftharpoons A(solv) + H^+(solv)$$

the corresponding equilibrium constant is:

$$K_{AH^+} = \frac{a_{H^+}a_A}{a_{AH^+}} = \frac{a_{H^+}[A]}{[AH^+]} \cdot \frac{\gamma_A}{\gamma_{AH^+}} \tag{3.1}$$

or in logarithmic form:

$$pK_{AH^+} = -\log a_{H^+} + \log I_A - \log\left(\frac{\gamma_A}{\gamma_{AH^+}}\right) \tag{3.2}$$

where the ionization ratio, $I_A = [AH^+]/[A]$. Rearranging:

$$-\log(a_{H^+}) = pK_{AH^+} - \log I_A + \log\left(\frac{\gamma_A}{\gamma_{AH^+}}\right) \tag{3.3}$$

it is apparent that the problem of determining $-\log a_{H^+}$, once the ionization ratio and pK_{AH^+} are known, lies in the behaviour of the last term, containing the activity coefficients. Hammett's method depends on the assumption that, for a series of progressively weaker bases used to ascend the scale by overlapping their ranges of measurable ionization ratios, the activity coefficient ratios for two bases A and B in the same solution are equal. (Cox and Yates 1983 describe this assumption as the 'zero-order approximation'.) For the second base, B,

$$-\log(a_{\mathrm{H^+}}) = \mathrm{pK_{BH^+}} - \log I_{\mathrm{B}} + \log\left(\frac{\gamma_{\mathrm{B}}}{\gamma_{\mathrm{BH^+}}}\right) \tag{3.4}$$

and, by subtraction:

$$\mathrm{pK_{AH^+}} - \mathrm{pK_{BH^+}} - \log I_{\mathrm{A}} + \log I_{\mathrm{B}} = -\log\left(\frac{\gamma_{\mathrm{A}}\gamma_{\mathrm{BH^+}}}{\gamma_{\mathrm{B}}\gamma_{\mathrm{AH^+}}}\right) = 0 \tag{3.5}$$

so $\mathrm{pK_{BH^+}}$ can be determined, and the scale continued as far as the ionization ratio of B can be measured. Hammett's H_0 scale depends on the use of a series of aromatic amines, of progressively weaker basicity. He defines:

$$H_0 = \mathrm{pK_{AH^+}} - I_{\mathrm{A}} = -\log[\mathrm{H^+}] - \log\left(\frac{\gamma_{\mathrm{H^+}}\gamma_{\mathrm{A}}}{\gamma_{\mathrm{AH^+}}}\right) \tag{3.6}$$

To extend this function to higher levels of acidity requires a succession of suitable weak bases. Each should have a $\mathrm{pK_{AH^+}}$ value close enough to that of its predecessor, that an overlapping range of acidity exists wherein the ratios I_{A} for both can be measured spectroscopically. It has been determined over a large range of acidities, from dilute aqueous solutions into the superacid range.

The foregoing assumes that A and B are of similar structure. If they are not, the assumption fails. Consider A and B as representing bases of two different structural classes. Most obviously, if the charges on the bases are different, the interionic parts of the activity coefficient ratios will differ. Using eqn 2.22 as a rough guide, if the interionic part of the activity coefficient of a univalent ion is give by γ_1^e then $\log(\gamma_z^e) = z^2\log(\gamma_1^e)$, and:

$$-\log\Gamma_z^e = -\log\left(\frac{\gamma_{\mathrm{H^+}}^e\gamma_{\mathrm{A^z}}^e}{\gamma_{\mathrm{AH^{Z+1}}}^e}\right) \approx ((z+1)^2 - 1 - z^2)\log(\gamma_1^e) = 2z\log(\gamma_1^e) \tag{3.7}$$

where z is the charge on the base, corresponding to the subscript on H_0, H_+, or H_-.

Even if A and B bear the same charge but are of different structure (e.g. a primary aromatic amine and an amide), the interactions of the bases and their protonated forms with solvent will differ, and the assumption still fails. Let A and B represent classes of bases of different structure. A scale dependent on bases of the B class can be established, but it will be different from the scale derived from class A, owing to differences in the mode and strength of solvation of the protonated and unprotonated species belonging to the A and B classes. Hammett (1970: 274) suggests, 'There is no reason except the large accumulation of data which accompanies its seniority in time for preferring H_0 as a measure of acidity over any of the other acidity functions.' This is still true, and must remain so until some means exists of determining the degree of solvation of the protonated and unprotonated forms of each of the indicator bases, perhaps by theoretical calculations of the sort mentioned in Chapter 5.

To reconcile the various scales several 'first-order' approximations have been proposed. Bunnett and Olsen (1966, and references in Cox 2000) assume that the logarithm of the activity-coefficient ratio for a base of class B is not equal to that for a Hammett base, A, but is proportional to it, that is,

$$\log \Gamma_{\mathrm{B}} = \log\left(\frac{\gamma_{\mathrm{H^+}}\gamma_{\mathrm{B}}}{\gamma_{\mathrm{BH^+}}}\right) = (1-\phi_{\mathrm{e}})\log\left(\frac{\gamma_{\mathrm{H^+}}\gamma_{\mathrm{A}}}{\gamma_{\mathrm{BH^+}}}\right) = (1-\phi_{\mathrm{e}})\log\Gamma_{\mathrm{A}} \quad (3.8)$$

So

$$H_{\mathrm{B}} + \log[\mathrm{H^+}] = -\log\Gamma_{\mathrm{B}} = -(1-\phi_{\mathrm{e}})\log\Gamma_{\mathrm{A}} = (1-\phi_{\mathrm{e}})(H_0 + \log[\mathrm{H^+}]) \quad (3.9)$$

and the B scale can be related to the A scale (H_0) provided the slope parameter ϕ_e can be determined. This can be done by plotting $\log I + H_0$ versus $H_0 + \log[\mathrm{H^+}]$, for by eqn 3.10, this plot gives the values of $\mathrm{pK_{BH^+}}$ (from the intercept) and ϕ_e (from the slope).

$$\log I + H_0 = \mathrm{pK_{BH^+}} + \phi_{\mathrm{e}}(\log[\mathrm{H^+}] + H_0) \quad (3.10)$$

An alternative method of reconciling the various scales is the *excess acidity method*, developed by Marziano *et al.* (1973, 1977) and confirmed and extended by Cox and Yates (1981, 1983; Cox 1987, 2000). Cox defines X as the excess acidity over the stoichiometric acid concentration, writing:

$$\log I - \log[\mathrm{H^+}] = m^*X + \mathrm{pK_{BH^+}} \quad (3.11)$$

for any indicator base B. Here m^*, the slope parameter, corresponds to $1 - \phi_e$. The data for many bases, in aqueous HCl (<40%), $HClO_4$ (<80%) and H_2SO_4 (<99.5%) have been correlated by computer (Cox and Yates 1978), and the results summarized by two empirical equations, eqn 3.12:

$$X = a_1(z-1) + a_2(z^2-1) + a_3(z^3-1) + \cdots \quad (3.12)$$

where $z = \text{antilog}(\text{wt\%}/100)$ for H_2SO_4 or $\text{antilog}(\text{wt\%}/80)$ for $HClO_4$, and the simpler eqn 3.13 for HCl:

$$X = a_1(\text{wt\%}) + a_2(\text{wt\%})^2 + a_3(\text{wt\%})^3 \quad (3.13)$$

Table 3.2 lists the coefficients for use in these equations to calculate the excess acidity, X. Figure 3.1 shows the excess acidity of solutions of sulphuric,

Table 3.2 Polynomial coefficients giving the excess acidity X as a function of wt% acid for aqueous sulphuric, perchloric, and hydrochloric acids (Cox and Yates 1978)

Coefficient	**H_2SO_4 (eqn 3.12)**	**$HClO_4$ (eqn 3.12)**	**HCl (eqn 3.13)**
a_1	−1.2192412	−0.74507718	0.0527767
a_2	1.7421259	1.0091461	0.00190497
a_3	−0.62972385	−0.30591601	−0.0000197423
a_4	0.11637637	0.049738522	—
a_5	−0.010456662	−0.0040517065	—
a_6	0.00036118026	0.00012855227	—

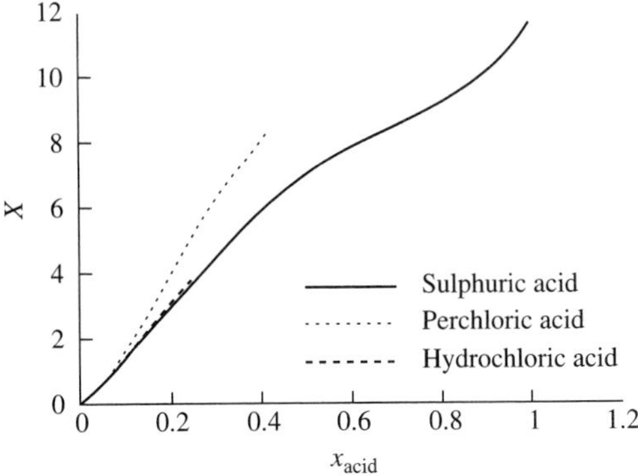

Fig. 3.1 Excess acidity, X, in aqueous perchloric, sulphuric and hydrochloric acid solutions versus mole fraction of acid. After Cox 2000, with permission.

perchloric and hydrochloric acids calculated from these equations, plotted against the mole fraction of each acid. The greater intrinsic strength of perchloric acid is apparent at mole fractions above 0.1.

Cox (2000) shows how the concept of excess acidity can be used in kinetic studies as an aid in assigning mechanisms. For example, data (Bell *et al.* 1956; Cox 2000) for the acid-catalysed depolymerization of the cyclic trimer of formaldehyde (trioxane) in several media (aqueous sulphuric, perchloric, and hydrochloric acids), were plotted as $\log k_\psi - \log C_{H^+}$ versus X (k_ψ represents the observed pseudo-first-order rate constant). The plots were highly linear over two orders of magnitude of X, with a slope of 1.091 ± 0.010. This was shown to be consistent with the *A-1 mechanism*:

$$S + H^+ \underset{\text{fast}}{\overset{K_{SH^+}}{\rightleftharpoons}} SH^+ \underset{\text{slow}}{\overset{k_1}{\longrightarrow}} \text{Products}$$

$$\mathbf{1} + H^+ \rightleftharpoons \mathbf{2}$$

Here S represents trioxane (**1**) and K_{SH^+} is the reciprocal of the (large) acid dissociation constant of the protonated trioxane (**2**).

In contrast, in the *A-2 mechanism*, the species SH^+ requires the assistance of a nucleophile, usually from the solvent, in the rate-determining step:

$$S + H^+ \underset{\text{fast}}{\overset{K_{SH^+}}{\rightleftharpoons}} SH^+ + Nu: \underset{\text{slow}}{\overset{k_2}{\longrightarrow}} \text{Products}$$

An example is the enolization of acetone in aqueous acid:

$$CH_3C(=O)\text{–}CH_3 + H_3O^+ \overset{K}{\rightleftharpoons} CH_3C(=HO^+)\text{–}CH_3 + H_2O$$

$$CH_3C^+(=HO^+)\text{–}CH_3 + H_2O \overset{k_2}{\longrightarrow} CH_3C(HO) = CH_2 + H_3O^+$$

The *A-2* mechanism can be distinguished from *A-1* by curvature in the log $k_\psi - \log C_{H^+}$ versus X plot (Cox 2000).

3.6. General acid/base catalysis and the kinetic solvent isotope effect

Reference was made above to involvement of the solvent in a reaction as catalyst. In a protic solvent reaction may be catalysed by the lyonium ion only (the hydronium ion, in water). This is specific hydrogen-ion catalysis. On the other hand, the reaction may be catalysed by any acidic species present in the solution (general acid catalysis). The solvent molecule itself may be a catalyst. Base catalysis, similarly, may be specific or general. For instance, the apparent rate constant for a reaction in aqueous acetate buffer solution might be represented by:

$$k_{obs} = k_0 + k_w[H_2O] + k_{H^+}[H_3O^+] + k_{OH^-}[OH^-] + k_a[HOAc] + k_b[AcO^-]$$

where the first term represents the 'uncatalysed' rate constant, the second the constant for the rate catalysed by water, acting as acid or as base, and the other terms as suggested by their subscripts. Disentangling these contributions requires rate measurements over ranges of pH with different concentrations of buffer components (Bell 1941, 1973; Jencks 1969).

If a reacting species with the natural isotopic composition is changed by substitution of one or more atoms by a less-common (usually heavier) isotope, a *kinetic isotope effect* (KIE) is observed. Simply viewed, the effect may be related to the difference between the zero-point energies (ZPEs) of the vibrational mode principally involving the substituted atom, in the initial and transition states. Assuming harmonic motion, the classical frequency of vibration of a bond between atoms of masses m_1 and m_2, with force constant k is given by eqn. 3.6.2, in which μ is the reduced mass, $m_1m_2/(m_1+m_2)$.

$$\nu = (1/2\pi)\sqrt{k/\mu} \tag{3.6.2}$$

The zero-point energy of vibration (ZPE) is then given by eqn. 3.6.3.

$$E_0 = \frac{1}{2}h\nu \tag{3.6.3}$$

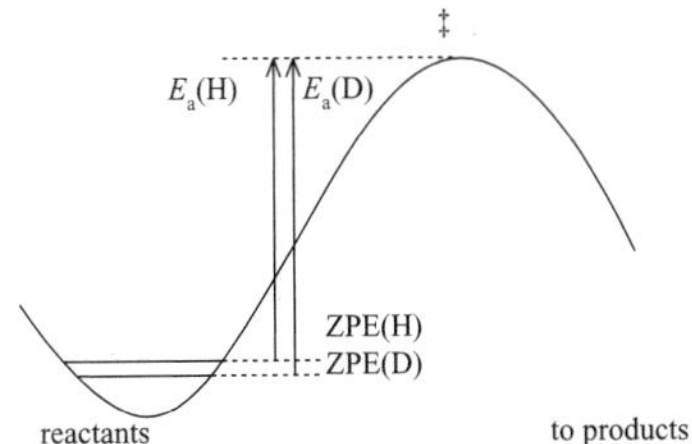

Fig. 3.2 Energy profile illustrating the effect of zero-point energy of the initial state only, on the activation energy for carbon-deuterium vs carbon-hydrogen bond scission, leading to a primary isotope effect. (Redrawn after Buncel and Dust 2000, with permission.)

To a good approximation, the Morse curve of potential energy vs interatomic distance is the same for different isotopic masses, so the force constant will also be essentially the same. If the atom being substituted is attached to a substantially heavier atom, i.e., if $m_1 \ll m_2$ then $\mu \approx m_1$. The relative difference in the square roots of the masses of isotopes is thus important, so the tritium isotope effect is potentially the largest, closely followed by deuterium. Heavy atom KIEs, while also smaller in magnitude, are nonetheless readily measurable and can be highly informative of reaction mechanism (Buncel and Saunders, 1992). Reactions in which the rate-determining step involves breaking the bond to the substituted atom are reduced in rate by substitution of a heavier isotope. This is the normal, *primary kinetic isotope effect.*

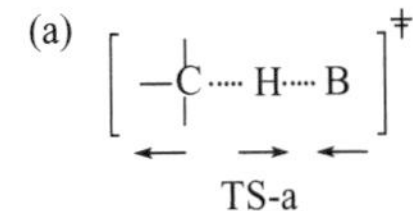

Fig. 3.3 Linear transition state: (a) antisymmetric stretch, (b) symmetric stretch.

Consider a reaction step in which a C–H (or C–D) bond is broken, and H is transferred from a molecule R–H to a base B. The reaction profile for the simple model is represented in Figure 3.2. Only the C-H stretch in the initial state is considered; the transition state (TS) is ignored. The vibrational energy of the C–D stretch lies below that for C–H by the difference in their ZPE's, leading to a factor approximately $1/\sqrt{2}$, i.e., 0.71. Estimating the force constant from typical R–H stretching frequencies observed in IR spectra around 3000 cm^{-1}, we can estimate the rate constant ratio $k_H/k_D \approx 7$ at 25°C, as a maximum value.

Other modes, in either the initial or transition states, can influence the outcome. To consider the TS first: if it is linear (Figure 3.3), there are two modes involving the C–H and H–B bond distances. The antisymmetric stretch (Figure 3.3a) corresponds to motion along the reaction coordinate, and is not associated with a ZPE. In the symmetric stretch (Figure 3.3b), the proton is nearly stationary (unless the masses of the C and B atoms are very different or the proton is off-centre), so this mode may still be neglected. A non-linear transition state, however, will have a symmetric stretch in which the proton moves, so the difference between the ZPE's for the TS's containing D and H will not be negligible, which will diminish the KIE. Inclusion of this and other modes whose ZPE's are altered by isotopic substitution in the initial state and in the TS (Figure 3.4) can lead to KIE's differing from the value, 7, derived from the simple model.

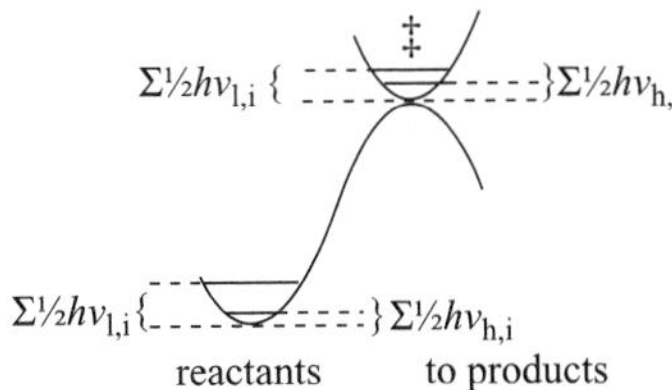

Fig. 3.4 A general energy profile that illustrates the origin of the kinetic isotope effect in terms of the zero-point energies of both the ground and transition (‡) states. The ZPE for the heavy isotope (subscript **h**) and the light isotope (subscript **l**) arise from the summation of the various vibrational modes, **i**. (Redrawn from Buncel and Dust 2000, with permission.)

Secondary isotope effects, which are generally smaller, are observed if the substituted atom is not directly involved. Several extensive theoretical treatments of kinetic isotope effects are available to the reader (Buncel and Dust 2003; Melander and Saunders, 1980; Collins and Bowman, 1970; see also the series, *Isotopes in Organic Chemistry*, Buncel and Saunders, Eds, 1992, and earlier volumes in the series, Buncel and Lee, eds.)

The *kinetic solvent isotope effect* (KSIE) is almost exclusively studied in terms of deuterium effects, because it arises naturally in the context of studies of general acid/base catalysis (Alvarez and Schowen, 1987; Kresge *et al.*, 1987). Smith and March (2001, p. 299) as well as Carroll (1998, p. 365) briefly describe the factors that may lead to KSIEs, including direct participation of the solvent as a reactant, exchange of deuterium between solvent and substrate before reaction, and differential solvation of reactants and transition state. The largest KSIE is expected when the rate-limiting step of a reaction involves breaking an O–H bond of the solvent, but two or all three factors may be simultaneously active. Alvarez and Schowen (1987), Kresge *et al.* (1987), and More O'Ferrall *et al.* (1971) discuss the separation of these factors. See also 'Secondary and Solvent Isotope Effects' vol. 7 in the series *Isotopes in Organic Chemistry*, Buncel and Lee, Eds. (1987).

There is currently much activity in the field of kinetic solvent isotope effects. A search using this phrase yielded 118 references to work on their use in elucidating a large variety of reaction mechanisms, nearly half in the last decade, ranging from elucidation of the SN2 process (Fang *et al.* 1998) to electron transfer in DNA duplexes (Shafirovitch *et al.* 2001). Nineteen countries were represented: see, e.g., Oh *et al.* (2002), Wood *et al.* (2002), Blagoeva *et al.* (2001), Koo, *et al.* (2001).

3.7 Lewis acids and bases

Lewis bases are species that can provide a pair of electrons to be shared with another species. Lewis bases are also Brønsted bases, for a proton can be the 'other species'. A Lewis acid, which is a species that can receive a share in a pair of electrons, cannot act directly as a Brønsted acid, but it can manifest acidity in an amphiprotic solvent by forming an adduct with a solvent molecule which is then a stronger acid than the solvent, for example

$$Al^{3+} + 7H_2O \rightarrow [Al(OH_2)_6]^{3+} + H_2O \rightleftharpoons [Al(OH_2)_5OH]^{2+} + H_3O^+$$

Thus very strong Lewis acids cannot exist in appreciable concentration in an amphiprotic solvent (nor can very strong Brønsted acids, owing to the levelling effect).

Various measures of Lewis base strength have been proposed. One such is the *donor number* (DN; Gutmann and Wychera 1966; Gutmann 1968), defined as the negative of the enthalpy of reaction of the base with the strong Lewis acid, antimony pentachloride, in dilute solution in 1,2-dichloroethane. This solvent was chosen as it is nearly (though not quite) non-basic. In fact, as Gutmann (1968: 19) points out, as well as the binding effect of the unshared electron pair, the donor number includes any dipole–dipole or ion–dipole contributions, and steric effects. It can be used to rationalize solubilities both of (Lewis) acids in basic solvents and of bases in acidic solvents, and differences in reactions of certain substances in different solvents. For instance, Gutmann (1968: 148) cites work by Mikhailov and coworkers showing that boron(III) chloride can donate a chloride ion to iron(III) chloride in diethyl ether (DN = 19.2) or in tetrahydrofuran (DN = 20.0) according to the reaction:

$$BCl_3 + FeCl_3 + 2(C_2H_5)_2O \rightleftharpoons [Cl_2B(O(C_2H_5)_2)_2]^+[FeCl_4]^-$$

but not in phosphorus oxychloride (DN = 11.7) and only to a certain extent in phenylphosphonic dichloride (DN = 18.5). Stabilization of $[BCl_2]^+$ requires a certain degree of donor strength on the part of the coordinating solvent. The concept of levelling, discussed above, is also used by Gutmann (1968: 168) in considering the effect of the coordinating power of the solvent, as measured by the donor number, toward transition-metal complex formation. In solvents of low donor number, such as nitromethane (DN = 2.7), reactions of the type exemplified by:

$$Fe^{3+}(S) + 4X^- \rightleftharpoons [FeX_4]^-$$

where X^- may be any of I^-, Br^-, Cl^-, N_3^-, or NCS^-, are all nearly complete when the reactants are present in stoichiometric amounts. Such reactions are thus levelled in this solvent. On the other hand, in dimethyl sulphoxide (DN = 29.9) $FeCl_3$ is completely solvolysed to $Fe(dmso)_n^{3+}$ and Cl^-, though the tetra-azido and hexathiocyanato complexes of Fe^{3+} are stable. DMSO is differentiating in this sense.

3.8 Hard and soft acids and bases ('HSAB')

Ahrland *et al.* (1958) introduced the a/b classification of metal cations. *Class* (*a*) contains those cations that form their most stable complexes with the ligand atoms of the first member of each group of the Periodic Table, that is, with F, O, or N rather than Cl, S, or P. *Class* (*b*) contains those ions that display the reverse order of stability. The reversal of the order of affinity of the halide ions as bases for the two acids Al^{3+} ($I^- \ll F^-$) and Hg^{2+} ($I^- \gg F^-$) is an example. Pearson (1963, 1988), in an attempt to account for these and related observations, noted that the small, highly charged ion Al^{3+} was most tightly bound to the small ion F^-, and the larger, less charged ion Hg^{2+} best bound the larger I^-, in keeping with Fajans's (1923) rules. He described as *hard*, acids in which the vacant orbital that accepts an electron pair is (when occupied) low-lying and relatively unpolarizable, and bases in which the donatable pair is similarly in a small, low-lying orbital and relatively unpolarizable. *Soft* acids and bases have relevant orbitals that are not so low in energy, and are relatively large and polarizable. Overlap is greatest and the interaction strongest when the orbitals involved are most closely similar in size and energy. Thus *hard acids react most strongly with hard bases, and soft with soft.* The presence of soft ligands tends to induce softness in a central atom: BF_3 is hard, but $B(CH_3)_3$ is borderline. In ligand reorganization reactions of the type:

$$MA_n + MB_n \rightleftharpoons MA_jB_k + MA_kB_j$$

(where $j+k=n$) if A and B differ in softness, the ligands might be expected to show discrimination[2] as defined in Chapter 2. Such cases have been reported, for example, the system $Si[N(CH_3)_2]_4$, $SiCl_4$ (Basolo and Pearson 1967: 445). Clark and Brimm (1965), however, found the distribution of the ligands PF_3 and CO among the five possible compounds resulting from the equilibration of a mixture of $Ni(PF_3)_4$ and $Ni(CO)_4$ to be essentially random, though PF_3 and CO are by no means equally soft. The stability of methylmercuric chloride towards disproportionation into $HgCl_2$ and $Hg(CH_3)_2$ (Buncel *et al.* 1986), an example of *antidiscrimination*, is explained by invoking subtleties of bond hybridization, implying competition between d orbitals on two chlorines for the same mercury orbital. *Discrimination* is favoured by difference in softness of the ligands, and disfavoured by great difference in their base strength (Pearson 1973: 51, 77). Considering, hypothetically, the halogenated methanes as formed from C^{4+} (hard), H^- (soft), F^- (hard) and Cl^- (intermediate), the HSAB theory would predict a greater degree of discrimination in the CH_nF_{4-n} system than in the CH_nCl_{4-n} system. The reactions are of course immeasurably slow, but the equilibrium constant calculated from Gibbs energies of formation for the reaction:

$$2CH_2X_2 \rightleftharpoons CH_4 + CX_4$$

with X=F is 8.5×10^{15}, and with X=Cl is 6.1×10^{-4}. The constant corresponding to unperturbed statistics is 0.028, so substantial discrimination exists in the fluoromethanes. The chloromethanes show a degree of antidiscrimination, perhaps merely due to crowding of the chlorines in CCl_4. The absence of d orbitals of low energy on carbon makes the argument using chlorine d orbitals less tenable.

3.9 Scales of hardness/softness

Edwards (1954, 1956), in a study of Lewis acid–base reactions involving metal cations and anionic ligands, showed that the equilibrium constants can be correlated by eqn 3.14:

$$\log\left(\frac{K}{K_0}\right) = \alpha E_n + \beta H \qquad (3.14)$$

in which the metal ions are characterized by a pair of parameters α and β, and the ligands by E_n and H.

[2] Jørgensen (1964) used the term 'symbiosis' for this effect, opposite in meaning to its sense in biology.

The ligand parameters are derived from independent properties of the ions Y^-: E_n from the standard oxidation potential for the reaction

$$Y^- = \tfrac{1}{2}Y_2 + e^-$$

and H from pK for the ionization of HY.

Figure 3.5 shows a plot of the values of α and β for 17 metal ions (Yingst and McDaniel 1967, cited by Jolly 1970: 53). Superimposed is a suggestion of a new coordinate system (effectively a rotation of the original coordinate system), *hardness* and *strength*. Except for the point for Zn^{2+}, which is classified by Pearson as 'borderline', but which nearly coincides with the hard Mg^{2+}, the *zero hardness* line (the 'borderline') divides the hard from the soft ions. The hardness coordinate shown is approximately represented by eqn 3.15;

$$H_+ = 0.623 - 0.465\alpha + 1.838\beta \tag{3.15}$$

Reactions of the type

$$A + :B \longrightarrow AB$$

where A and :B are a Lewis acid and base have been studied by calorimetry (in a poorly-coordinating solvent) by Drago and Wayland (1965). They showed that the enthalpy of reaction between uncharged species can be expressed by eqn 3.16.

$$-\Delta H = E_A E_B + C_A C_B \tag{3.16}$$

Acid and base strength are represented not by a single number for each, but rather two numbers, E_A and C_A, to characterize the acid, and similarly E_B and C_B to characterize the base. They interpret the E's as measuring the tendency

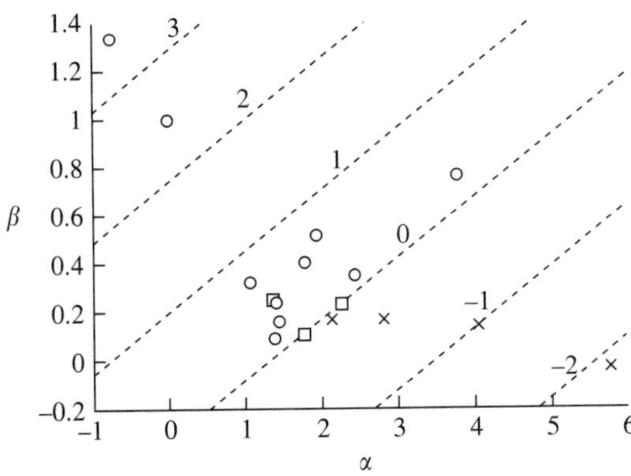

Fig. 3.5 Comparison of Edwards's acidity parameters with Pearson's hard/soft classification. Circles represent hard acids, crosses soft, squares borderline. Dashed lines are for the indicated values of a hardness coordinate (arbitrary scale), defined by: $H_+ = 0.623 - 0.465\alpha + 1.838\beta$.

to electrostatic bonding and the C's as measuring the tendency to covalent bonding. Some values of the parameters are tabulated in the Appendix. As with the Pearson and Edwards descriptions, discussed above, the Pearson and Drago descriptions are also related somewhat like two coordinate systems in a plane, differing in orientation, and with some displacement of the origin (see Fig. 3.6).

It is probably too much to expect that a single scale of softness (or hardness) could be established for all species. Drago *et al.* (1991), in considering reactions in which one or both reacting species is charged, find that it is necessary to introduce an additional term $R_A T_B$ to eqn 3.16, to account for electron density transferred. This complicates the interpretation of softness as one of a pair of rotated coordinates. It may be necessary to define a set of softness scales for acids, S_ν^A, where ν is the charge on the species or at the active site, and a similar set, S_ν^B, for bases. Huheey *et al.* (1993: 349) present a table of values of equilibrium constants (as pK) for a variety of neutral and anionic bases reacting with the hard acid H_3O^+ and the soft acid CH_3Hg^+. There are clear cases where the strength of the base can be related to the sum of the pK's and the softness to their difference, but the presence in the list of bases of monovalent and divalent anions, of simple neutrals like ammonia,

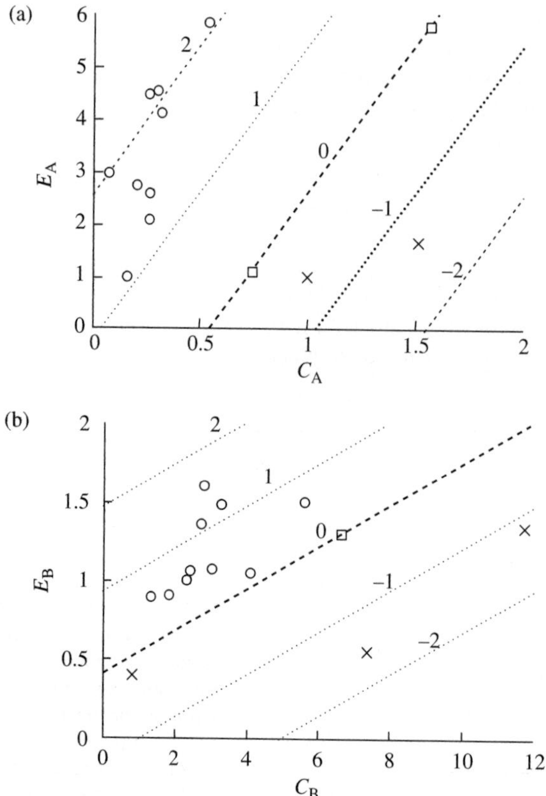

Fig. 3.6 Comparison of Drago's acid (a) and base (b) parameters with Pearson's hard/soft classification. Crosses represent soft acids or bases, circles hard, and squares borderline. The dashed lines are for the indicated values of hardness (arbitrary scales).

with a single basic site, and species like SCN^-, dimethyl sulphoxide, and $Et_2PCH_2CH_2OH$, which have two sites of widely differing softness, make the interpretation difficult.

Marcus (1987) defines the (base) softness parameter μ for a solvent as 'the difference between the mean of the Gibbs free energies of transfer of sodium and potassium ions from water to a given solvent and the corresponding quantity for silver ion (/kJ mol^{-1}), divided by 100.' He points out that μ is uncorrelated with the β scale of solvent basicity (Kamlet and Taft 1976), which is a measure of hydrogen-bond acceptor strength, 'hard' basicity. It might be expected that μ might show some correlation with the volume refraction, $R_v = (n^2 - 1)/(n^2 + 2)$, since softness is considered to be related to polarizability. Multilinear correlation of μ with R_v and Swain's *acity* and *basity* parameters, A_j and B_j, in the form:

$$\mu = C_0 + C_1 R_\nu + C_2 A_j + C_3 B_j \tag{3.17}$$

yielded values of constants listed in Table 3.3. Softness is not a bulk property, but a molecular one, or more precisely a property associated with the acidic or basic site within a molecule. A close correlation with R_v (a bulk property) or with $[R]$, the molar refraction (Lorentz–Lorenz) should not be expected. Nevertheless, both were tried. Clearly the correlation coefficients with A_j and B_j are zero in either form. The correlation with $[R]$ is only slightly better (relative standard deviation of slope = 0.60) than with R_v (0.41).

Four solvents (ammonia, acetonitrile, pyridine, and triethylamine) did not fit either correlation, μ for these lying some 0.2–0.5 units above the trend of the remainder. That they are all nitrogen compounds that have a particular affinity for silver ion may be significant. In a similar correlation of μ with Drago's C_b and E_b for the very few solvents for which all three quantities are known, ammonia lay 0.6 units above the trend (two other nitrogen compounds, *N,N*-dimethylformamide and hexamethylphosphoramide, are not apparently anomalous). Softness is also expected to be related to the energy gap between the *highest occupied* and *lowest unoccupied molecular orbitals* (HOMO and LUMO). These are also molecular properties, so their relevance to the local properties at acid or base sites varies with the complexity of the molecule. For atoms, or for small molecules with a central atom and simple ligands, softness should be strongly correlated with the reciprocal of the

Table 3.3 Multilinear correlation of the Marcus softness parameter μ with either the volume refraction R_v or molar refraction $[R]$[a] and with Swain's parameters A_j and B_j

	C_0	C_1	C_2	C_3
Using R_v	-0.4 ± 0.4	2.3 ± 1.4	-0.07 ± 0.19	-0.04 ± 0.22
Using $[R]$	-0.08 ± 0.10	0.0094 ± 0.0045	0.05 ± 0.18	-0.01 ± 0.22

[a] $R_v = (n^2 - 1)/(n^2 + 2)$; $[R] = R_v M/\rho$

HOMO–LUMO gap. The comparison of several base-softness scales proposed by Chen *et al.* (2000), and others, is discussed in Chapter 4.

3.10 Acids and bases in reactive aprotic solvents

The term *autoionization* (but not *autoprotolysis*) may be applied to reactions of certain aprotic solvents, for example

$$2SbCl_5 = SbCl_4^+ + SbCl_6^-$$

Here the ion transferred is Cl^-, rather than H^+. Addition of a strong Lewis acid to such a solvent will result in the appearance in solution of the characteristic positive ion of the solvent. Addition of a base will yield the negative ion. Again, aprotic solvents may be levelling towards acids stronger than the characteristic acid ion of the solvent, or, *mutatis mutandis*, towards bases. It will be seen that this is consistent with the picture based on the proton.

Generally, bases are recognized by their ability to provide negative charge in the form of actual negative ions or of electron pairs to be shared. Acids are recognized by their ability to provide positive charge or to accept shared electron pairs. In the context of reaction kinetics and mechanisms, a base is termed a *nucleophile*, and a Lewis acid an *electrophile*.

3.11 Extremes of acidity and basicity

J. B. Conant (Hall and Conant 1927) recognized that certain systems, such as Friedel–Crafts catalysts, can have acidities greater than that of 100% H_2SO_4 ($H_0 = -12$). Gillespie (1971, 1973) called these *superacids*. The strongest known (Gillespie and Peel 1972) simple Brønsted acid is fluorosulphonic acid, FSO_3H, with $H_0 = -15.1$ for the neat acid, closely followed by trifluoromethane-sulfonic acid ('triflic acid') at −14.1. Even greater acidities can be achieved by the formation of complexes between certain Brønsted and Lewis acids. A solution of 1 molar SbF_5 in HF has been shown to have a value of H_0 beyond −22 (Fabre *et al.* 1982; Olah *et al.* 1985). These acid systems extend greatly the possibilities for acid catalysis, and make possible the preparation of solutions containing carbocations, stable enough for study.

The complementary term is *superbase*. Solvents that are sufficiently basic (e.g. ammonia), or at least weak enough acids (e.g. dimethyl sulphoxide) can support high levels of basicity. Jolly (1970: 104) shows a figure representing the effective pH ranges (water scale) possible in several solvents, extending to 30 in DMSO, and to 37 in ammonia. Amide ion in ammonia can deprotonate molecular hydrogen enough to catalyse the exchange reaction:

$$HD + NH_3 \rightleftharpoons H_2 + NH_2D$$

which has been considered as a basis for deuterium enrichment (Buncel and Symons 1986).

3.12 Oxidation and reduction

The strength of an oxidant or a reductant that can exist for an indefinite period in any given solvent is obviously limited by the susceptibility of the solvent to oxidation or reduction. One must, however, consider both the thermodynamic and the kinetic aspects of any possible reaction. For instance, in aqueous solution in the absence of catalysts, oxidants that should, thermodynamically, be able to drive the half-reaction

$$2H_2O = 4H^+ + O_2 + 4e^-$$

can nevertheless exist in solution as useful reagents: permanganate, for instance. Similarly, the reaction

$$2Cr^{2+}(aq) + 2H^+(aq) = 2Cr^{3+}(aq) + H_2(g) \quad K = 9 \times 10^6$$

is so slow that an aqueous solution containing chromous ion may be used as a very strong reductant, as long as oxygen is excluded.

Liquid ammonia dissolves the alkali metals without immediate reaction, and affords a strong reducing agent, as long as catalysts (chiefly finely-divided metals) for the reaction:

$$2NH_3 + 2e^-(am) \rightleftharpoons 2NH_2^- + H_2$$

are excluded.

Barthel and Gores (1994: 20) list 23 solvents, selected from several classes, that can sustain large ranges of electrochemical potential, depending on the added electrolyte. The largest range reported is 6.8 V (from −3.3 to +3.5 V on Pt versus Ag/Ag^+) for ethylene carbonate containing potassium hexafluorophosphate. Coetzee and Deshmukh (1990) warn that impurities, some difficult to remove, may seriously limit the ranges of oxidation/reduction potential, as well as of acidity or basicity.

3.13 Acidity/redox diagrams

Pourbaix (1949, 1963) diagrams are widely used, especially in geochemistry, to show the conditions of pH and reduction potential in aqueous media in which various species of an element are stable. Figure 3.7 shows such a diagram for Fe species in water at 25 °C. The ordinate is the reduction potential with respect to the standard hydrogen electrode. The solid lines correspond to equilibria between two species at equal concentrations or activities; the fields are labelled with the species that are stable within them.

Consider a half-reaction: $\mathbf{O} + m\mathrm{H}^+ + n\mathbf{e}^- = \mathbf{R}$, where **O** and **R** are the oxidized and reduced principal species. From the form of the Nernst equation, eqn 3.18, in which F is the Faraday constant,

$$\begin{aligned} E &= E_0 - \frac{RT}{2.303nF}\log\left(\frac{[\mathrm{R}]}{[\mathrm{O}][\mathrm{H}^+]^m}\right) \\ &= E_0 - \frac{RT}{2.303F}\frac{m}{n}\mathrm{pH} - \frac{RT}{2.303F}\log\left(\frac{[\mathrm{R}]}{[\mathrm{O}]}\right) \end{aligned} \tag{3.18}$$

it may be seen that the theoretical slope of the line along which the activities of **O** and **R** are equal is given by $-m/n$, the ratio of the numbers of protons to electrons transferred, times the factor $RT/2.303F$ (though some authors use, as ordinate, $\mathbf{pe} \equiv (2.303F/RT)\,E$, so the slope is just $-m/n$).

The concept can be extended to organic systems. Figure 3.8 shows a similar diagram for isobutane (symbolized as iC_4H) and derivatives in liquid HF containing SbF_5 or KF (Fabre *et al.* 1982). Here the ordinate is the potential with respect to the $Ag/AgSbF_6$ electrode, and the abscissa is pH(HF), that is, the pH on a scale with zero corresponding to a 1 M solution of strong acid in HF. The accessible range of pH is that spanned by 1 M strong acid (0) to 1 M strong base (13.7) in HF, corresponding approximately to H_0 from -22 to -8. In the figure 'iC_4' represents the *t*-butyl group $(CH_3)_3C$; '$iC_{4=}$' represents isobutene. In using any of these diagrams, one must be aware that they are constructed on the assumption that only specified species may be present, and only specified equilibria are considered (in Fig. 3.4, for instance, the species $Fe(OH)^{2+}$, $Fe(OH)_2^+$ and $Fe(OH)_4^-$ are omitted, and H_2FeO_4 is assumed to be a strong acid). With that caution, such a diagram may be used to predict the conditions of acidity and reduction potential in which a species is stable with respect to any of the reactions that were considered in the construction of the diagram. A species may be unreactive enough to remain present outside the predicted area of stability (most organic compounds are unstable to oxidation by the atmosphere, though the 'autooxidation' is usually immeasurably slow).

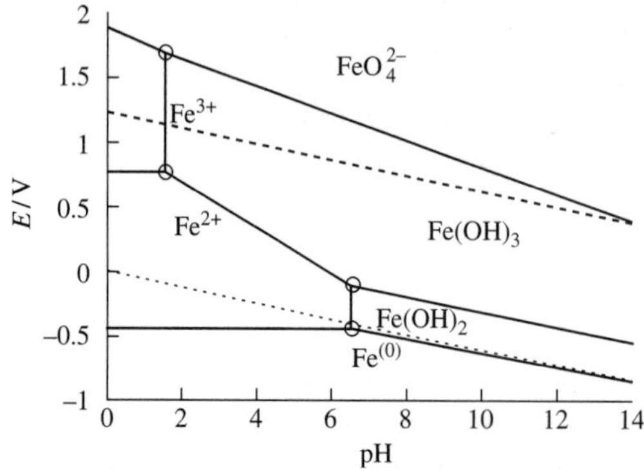

Fig 3.7 Pourbaix diagram for iron species in aqueous environment, 25 °C. Dashed lines indicate the thermodynamic thresholds for evolution of oxygen (upper) or hydrogen (lower).

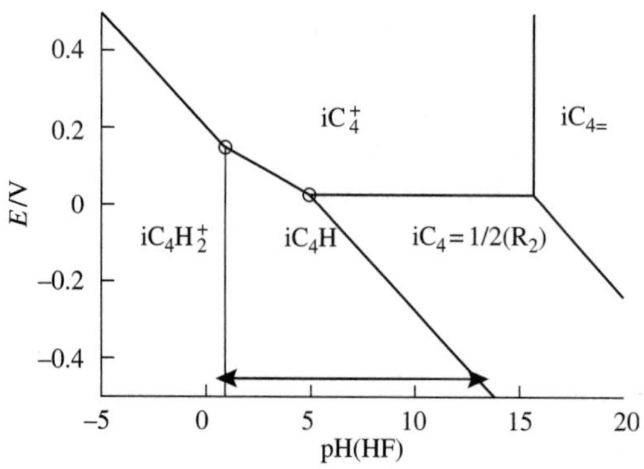

Fig. 3.8 Pourbaix diagram for species derived from isobutane in solution in HF. The neutral point in HF is at $pH = 6.9$, at which $H_0 \approx -15$. Addition of KF or SbF_5 serves to adjust pH to more basic or more acidic values, within the practical limits 1–13, approximately, a range of H_0 about -21 to -9 (double arrow). Redrawn after Fabre *et al.* (1982), with permission.

Trémillon (1971) describes the use of E versus $p(O^{2-})$ diagrams, comparable to E versus pH diagrams in the study of solutions in fused salts or fused alkali hydroxides. Similar considerations apply.

3.14 Unification of acid/base and redox concepts

It has been noted by Usanovich (1939) that properties of bases and acids are in a sense parallel to those of reducing and oxidizing agents. In clear cases it is possible to show that one can formulate an oxidation–reduction reaction as a total transfer of one or more electrons (commonly with such reactions in aqueous solution) or of an atom, and an acid–base reaction as a transfer of an ion. In some circumstances the distinction may be difficult to define. The acquisition of a fourth oxygen atom by sulphite looks very much like a Lewis neutralization:

$$:\ddot{O}: + :SO_3^- \rightarrow :\underset{\cdot\cdot}{\ddot{O}}: SO_3^-$$

with the oxygen atom acting as acid and sulphite as base. The solvated electron, present in the solution of sodium metal in liquid ammonia, can act both as a reducing agent and as a very strong base.

Usanovich took the step of combining the two concepts, and defining acids and bases in terms of all kinds of charge generation or transfer, calling 'acids' all species that donate positive charge or accept negative charge, whether by outright transfer or by sharing, and calling 'bases' all species that donate negative charge or accept positive charge. Few chemists are willing to follow him so far, as the distinction between acids and bases on the one hand and oxidizing and reducing agents on the other is important. Nevertheless, it is helpful in some cases to keep the parallel nature of these two concepts in mind.

Problems

3.1. The following reactions, carried out in water, have equilibrium constants greater than unity. Identify the acid species involved and sort as many of them as possible into descending order of strength.

(i) $NH_3 + H_2CO_3 \rightleftharpoons NH_4^+ + HCO_3^-$
(ii) $H_2CO_3 + CN^- \rightleftharpoons HCO_3^- + HCN$
(iii) $H_3PO_4 + CH_3COO^- \rightleftharpoons H_2PO_4^- + CH_3COOH$
(iv) $HSO_4^- + H_2PO_4^- \rightleftharpoons SO_4^{2-} + H_3PO_4$
(v) $H_3O^+ + SO_4^{2-} \rightleftharpoons H_2O + HSO_4^-$
(vi) $CH_3COOH + NH_3 \rightleftharpoons CH_3COO^- + NH_4^+$
(vii) $HCN + CO_3^{2-} \rightleftharpoons CN^- + HCO_3^-$

3.2. For ethanoic acid, $pK_a = 4.77$. To be consistent with this value, should the pK_a for H_3O^+ be taken as zero or as -1.74? Explain.

3.3. For the weak acid, hydrogenphthalate ion, HA^-, the aqueous dissociation constant may be written:

$$K = \frac{a_H + a_{A^{2-}}}{a_{AH^-}} = \frac{[H^+][A^{2-}]}{[AH^-]} \cdot \frac{f_{H^+} f_{A^{2-}}}{f_{AH^-}} = K_c \cdot \frac{f_{H^+} f_{A^{2-}}}{f_{AH^-}}$$

so the 'extended' Debye–Hückel equation (with a nod to Davies) for moderate ionic strengths suggests:

$$\ln K = \ln K_c + \text{Constant}\left(\frac{I^{1/2}}{1 + I^{1/2}}\right)$$

Determine (graphically or otherwise) the best value of K from the following data.

I (mol L^{-1})	0.0519	0.1035	0.1551	0.2066	0.323	0.5163
$K_c \times 10^6$	9.11	11.55	13.72	15.73	19.52	24.81

3.4. In (separate) aqueous sulphuric acid solutions of two Hammett bases, A and B, containing 1.00 μmol L^{-1} of the respective bases, the tabulated absorbances, due to the base or its conjugate acid, were observed. All readings were taken using a 1.00-cm cell. The value of pK_a for HA^+ was −0.053. Obtain an estimate of pK_a for HB^+, and of the excess acidity X in 1.3, 4.5, and 7.2 M H_2SO_4. Assume spectroscopic medium effects are absent.

	$C_A = 1$ μmol L^{-1}		$C_B = 1$ μmol L^{-1}	
C(acid)(mol L^{-1})	**$A(\lambda_1)$**	**$A(\lambda_2)$**	**$A(\lambda_3)$**	**$A(\lambda_4)$**
0.01	0.773	0.034	1.005	0.009
10	0.002	0.506	0.011	0.705
1.3	0.243	0.358	0.905	0.079
4.5	0.011	0.5	0.202	0.571
7.2	0.003	0.506	0.023	0.696

3.5. Azoxybenzene (**3**) is stable for indefinite periods in aqueous, alkaline or dilute acid media. In moderately concentrated sulphuric acid solution, however, it undergoes rearrangement to *p*-hydroxyazobenzene (**4**), a reaction known as the Wallach rearrangement, Equation (a):

(a)

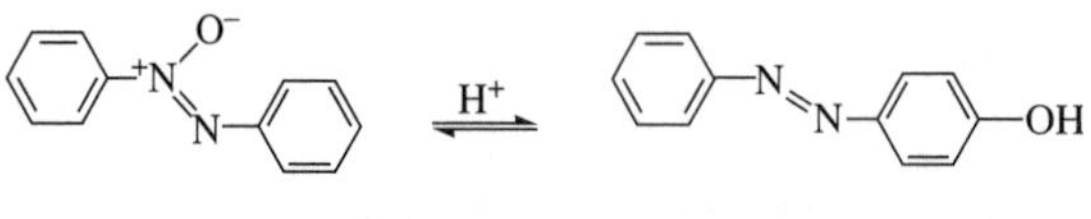

H^+

3 **4**

(b)

H^+

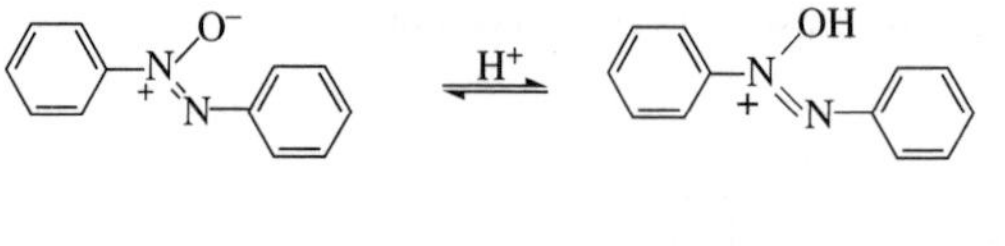

Table 3.4 Kinetic data for the rearrangement of azoxybenzene to 4-hydroxyazobenzene at 25 °C.

H_2SO_4 (wt%)	X[a]	$-H_0$[b]	α[c]	$10^5 k_\psi/s^{-1}$[d]
75.30	5.30	6.65	0.967	0.106
80.15	6.17	7.42	0.994	0.208
85.61	7.20	8.35	0.999	2.17
90.37	8.05	9.05	1.000	7.23
95.12	9.04	9.82	1.000	20.9
97.78	10.01	10.35	1.000	43.8
99.00	10.75	10.82	1.000	76.8
99.59	11.21	11.18	1.000	227
99.90	11.54	11.48	1.000	2310
99.99	11.56	11.90	1.000	4160

[a] Data from Buncel (2000).
[b] Data from Cox (1987).
[c] Fraction of azoxybenzene protonated, calculated using $pK_a = -5.15$ (Buncel and Lawton 1965).
[d] Pseudo-first-order rate constants as determined spectrophotometrically.

A kinetic study of this reaction in media of different acid concentrations (Buncel 2000) gave the results shown in Table 3.4, which records as well the corresponding X and H_0 values. Also given in Table 3.4 are the extents of protonation of azoxybenzene, according to eqn (b), which are calculated (Buncel 1975a) from the spectrophotometrically determined value, $pK_a = -5.15$, corresponding to 50% protonation of azoxybenzene in 65% H_2SO_4 ($H_0 = -5.15$). From these results, what can you deduce about the role of protonation of **3** in its ease of rearrangement, and hence on the mechanism of Reaction (a)?

4 Chemometrics: Empirical correlations of solvent effects

4.1 Linear free energy relationships

In the previous chapters we have considered the effects of physical properties (cohesive forces, polarity, polarizability) and chemical properties (chiefly acidity and basicity in their various manifestations) on equilibria and rates of reaction. The problem with theoretical approaches to such matters is that these properties never act alone, nor are they independent. For instance, in a series of solvents that are weakly acidic, the acidity, hydrogen-bonding ability, polarity, and solubility parameter may all be expected to increase more or less together.

Many attempts have been made to bypass the need to disentangle the parallel effects of such properties of solvent molecules, by noting the correlation of the effects of changing solvents on one phenomenon with another believed to be susceptible to similar influences. The phenomena studied may be chemical equilibria or reaction rates, but they may also be features of electromagnetic spectra: ultraviolet/visible, infrared, or nuclear magnetic resonance. By setting up a scale based on one phenomenon, usually a spectral frequency or the logarithm of an equilibrium or rate constant, one may then seek other cases where the data in suitable form, plotted against this scale as abscissa, yield a straight line. The slope of the line may then be interpreted as a measure of the susceptibility of the molecule or the reaction to changes in the particular solvent property on which the scale is based.

This approach is often termed *linear free-energy relationship* (LFER) (Hammett 1970; Chapman and Shorter 1972), because it seems to work best when the scales are based on the logarithms of equilibrium constants or rate constants, which are related to the Gibbs free energy of reaction (eqn 1.1), or of activation (eqn 1.18), or to changes in spectral frequencies, which are proportional to energy changes. The correlation of substituent effects on rates and equilibria, by Hammett (1937, 1970) is a familiar example. In the present context, the solvent properties that should be considered important are those that have been discussed in the foregoing chapters, namely general properties such as polarity, the solubility parameter and polarizability, and more specific properties such as hydrogen-bonding ability (whether as donor or acceptor) and Brønsted or Lewis acidity and basicity. One is then concerned with the correlation of these properties with solvent effects on other types of equilibria or on reaction rates. The term *chemometrics* (Wold and Sjöström 1982) is used to refer to all the various methods of treating chemical data through statistics: correlations of various kinds, and especially *principal component analysis*, discussed in Section 4.2.

Looking over the array of empirical parameters that have been derived by various authors (see references in Appendix Table A.2 and in Reichardt 1988, ch. 7) to correlate effects on reaction equilibria, rates or spectral frequencies, it appears that there are many effects of the solvent to be considered. The parameters can be divided into two broad categories. First are those that have no 'sign,' that is, they are in principle *symmetric* in their effect on cationic or anionic species or on molecules that have electron donor or acceptor properties. These are such as cohesive energy density or cohesive pressure (and its square root, the solubility parameter), internal pressure,[1] polarity, polarizability, refractive index, dielectric constant, and a number of empirical parameters based on particular equilibria, rates or spectral features. An assortment of these parameters is listed in Appendix Table A.2(a), with an indication of the experimental basis of each.

The second class contains *dual parameters*, which occur in pairs of complementary attributes, such as cationic and anionic charge, Lewis or Brønsted acidity and basicity (and refinements such as hard or soft acidity and basicity), electrophilicity and nucleophilicity, and hydrogen-bonding tendency as donor and as acceptor (Appendix Table A.2(b)). A number of the entries in the latter table are incomplete, in that only one of a possible pair of complementary parameters has been investigated. A table of values of most of the listed parameters for selected solvents forms Appendix Table A.3.

Many of the listed parameters are used in equations of the form of eqn 4.1:

$$Z = Z_0 + aA + bB + etc. \tag{4.1}$$

where Z represents the value of a property (usually the logarithm of a reaction rate or equilibrium constant) for a system of interest, Z_0 the value of Z for a reference system, the other capital letters represent parameters characteristic of the solvent, and lower-case letters coefficients of the corresponding parameters, characteristic of the particular system. An example is the Kamlet–Taft correlation: see Section 4.4.

As noted above, the solvent parameters are not independent. Clearly, if a reaction gives rise as a product or as an activated complex to a species that is cationic or has a site with localized positive charge, the reaction will be favoured by solvent properties including polarity, polarizability, basicity (whether hard or soft), and by tendencies to covalent or electrostatic interaction with vacant orbitals (i.e. nucleophilicity). Similarly, if the product or activated complex bears a localized negative charge, the reaction will still be favoured by polarity and polarizability, but also by acidity, and by the presence of vacant orbitals capable of receiving electron donation (electrophilicity).

Some other relationships are not so obvious: for instance, Dunn *et al.* (1984) found the boiling point and the molar refractivity of some halogenated

[1] The internal pressure is defined by $P_i = (\partial U/\partial V)_T$. It is of doubtful significance for our purposes; for water below 4 °C, for instance, it is negative.

hydrocarbons to be strongly correlated ($r = 0.92$). Hildebrand's δ and the polarizability are also correlated. In the simple case of non-polar, spherical molecules, the London theory of intermolecular forces relates the attractive force between two molecules to the product of their polarizabilities. These forces can also be related to the ionization potentials of the molecules, which are in turn closely related to the energies of the highest occupied molecular orbitals (HOMO). One would hope that by finding out to what extent all these parameters are interrelated, the number of distinct, independent variables could be much reduced. Many of the supposedly symmetric parameters in Appendix Table A.2(a) are also related to some of the dual parameters in Appendix Table A.2(b). The quantity $E_T(30)$, for instance, usually treated as one of the symmetrical parameters, contains a strong component of acidity (hydrogen-bond donation).

4.2 Correlations between empirical parameters and measurable solvent properties

One would like to be able to understand the effects of solvents upon reactions and spectra in terms of properties of molecules or bulk solvents such as:

- polarizability
- suitable functions of refractive index
- corresponding functions of dielectric constant
- dipole moment
- quadrupole moment
- Lewis acidity and basicity
- hardness and softness
- hydrogen-bond donor or acceptor strength

With this in view, statistical correlations have been undertaken by a number of authors, either pairwise (between two variables at a time) or between one, believed to be composite, and several others in multiple linear correlation, that is to say, seeking relationships of the form:

$$T = C_0 + C_1x + C_2y + C_3z + \cdots \tag{4.2}$$

where T is the composite variable, and $x, y, z, \ldots$ represent properties expected to contribute to the variance of T.

The parameters listed in Table 4.1 (a selection of the parameters listed in Appendix Tables A.2(a) and (b)) were subjected to linear correlation, two at a time. The results of the calculations are displayed in Table 4.2: the binary correlation coefficients r_{jk}, along with the numbers n_{jk} of solvents for which values of both parameters were available. Values of $|r_{jk}|$ below 0.2 suggest that the parameters j and k are essentially independent; values close to unity ($|r_{jk}| > 0.9$) suggest that the two measure essentially the same property. Intermediate values signify mixing of two or more. For example, E_T^N, $\beta_\mu^{1/2}$, Z, and A_j all show moderate cross-correlation. Any one of them could possibly be represented as a linear combination of all the others in this group.

Table 4.1 Parameters correlated. For definitions and references, see Appendix Tables A.2(a) and (b)

1	R_v	9	$A(^{14}N)$	17	A_j
2	Q^v	10	S	18	B_j
3	$\beta_\mu^{1/2}$	11	χ_R	19	α
4	δ_H	12	$\mathcal{S}$	20	β
5	E_T^N	13	W	21	AN
6	Z	14	Ω	22	DN
7	π^*	15	μ	22	E_b
8	π^*_{azo}	16	$\log K_{o/w}$	24	C_b

Table 4.2 Pairwise correlation coefficients between selected solvent parameters. *Continued on page 70*

No.	R_v	Q_v	$\beta_\mu^{1/2}$	δ_H	E_T^N	Z	π^*	π^*_{azo}	$A(^{14}N)$	S	χ_R	$\mathcal{S}$
	1	2	3	4	5	6	7	8	9	10	11	12
1	1	29	29	23	29	29	29	29	23	25	24	17
2	−0.496	1	29	23	29	29	29	29	23	25	24	17
3	−0.360	0.887	1	23	29	29	29	29	23	25	24	17
4	0.107	0.640	0.748	1	23	23	23	23	18	19	19	15
5	−0.476	0.839	0.658	0.821	1	29	29	29	23	25	24	17
6	−0.489	0.791	0.614	0.823	0.977	1	29	29	23	25	24	17
7	−0.002	0.715	0.787	0.770	0.595	0.545	1	29	23	25	24	17
8	−0.025	0.787	0.826	0.779	0.694	0.652	0.960	1	23	25	24	17
9	−0.445	0.675	0.517	0.679	0.938	0.932	0.546	0.657	1	21	20	15
10	−0.410	0.787	0.616	0.879	0.977	0.956	0.603	0.716	0.934	1	22	16
11	−0.060	−0.821	−0.689	−0.831	−0.860	−0.809	−0.783	−0.892	−0.849	−0.849	1	16
12	−0.001	0.615	0.718	0.808	0.453	0.328	0.969	0.904	0.254	0.404	−0.710	1
13	−0.168	0.883	0.551	0.968	0.962	0.966	0.633	0.754	0.890	0.948	−0.883	0.396
14	−0.593	0.761	0.641	0.865	0.934	0.976	0.598	0.718	0.872	0.912	0.772	0.243
15	0.572	−0.175	0.032	0.072	−0.379	−0.396	0.155	0.150	−0.379	−0.395	0.072	0.154
16	0.367	−0.800	−0.750	−0.725	−0.802	−0.791	−0.595	−0.652	−0.693	−0.801	0.726	−0.399
17	−0.451	0.686	0.444	0.741	0.948	0.939	0.455	0.549	0.969	0.976	−0.825	0.363
18	0.060	0.661	0.789	0.767	0.491	0.448	0.960	0.923	0.450	0.498	−0.741	0.992
19	−0.553	0.560	0.276	0.150	0.870	0.866	0.236	0.360	0.900	0.899	−0.529	0.073
20	−0.246	0.644	0.588	0.305	0.482	0.433	0.361	0.448	0.362	0.510	−0.480	0.177
21	−0.406	0.666	0.406	0.785	0.952	0.954	0.456	0.575	0.959	0.967	−0.765	0.163
22	0.340	0.293	0.326	0.076	0.149	0.061	0.361	0.437	0.164	0.086	−0.453	0.056
23	−0.244	0.542	0.526	0.078	0.206	0.236	0.260	0.333	0.120	0.048	−0.160	0.016
24	0.122	−0.273	−0.226	−0.304	−0.395	−0.432	−0.431	−0.344	−0.219	−0.473	0.287	−0.132

Entries above the diagonal represent the number of solvents $n_{i,j}$ for which both parameters were known.
Entries on and below the diagonal are correlation coefficients, defined by:

$$r_{i,j} = (n_{i,j}\Sigma x_i x_j - \Sigma x_i \Sigma x_j)/[(n_{i,j}\Sigma x_i^2 - (\Sigma x_i)^2)(n_{i,j}\Sigma x_j^2 - (\Sigma x_j)^2)]$$

To disentangle these cross-relationships, the statistical method of *principal component analysis* (PCA) (Eliasson *et al.* 1982; Thielemans and Massart 1985; Malinowski and Howery 1989; Johnson 1998) may be used. The purpose of PCA is to express the various properties as linear combinations of a set of new variables, chosen to be independent of each other, that is to say, they are orthogonal. Any of the original variables can be written as a linear combination of the orthogonal set, just as a vector in three-dimensional space can be written in terms of three vectors, mutually at right angles. Here, then we write for each property P_j

$$P_j = \sum a_{j,r} Z_r. \tag{4.3}$$

The number of the new variables, Z_r, is equal to the number of properties that were in the original set. It is a feature of PCA, however, that the *principal components* (PCs), Z_r are oriented and ordered so that the first, Z_1, covers the largest part of the variation of the original data, Z_2 the next largest, and so on. In favourable cases, the number of the new variables required to reproduce all the original data within experimental uncertainty, or at least to sufficient precision for their purpose, may be small. The sum in eqn 4.3, then, may

Table 4.2 *Continued from page 69*

W	Ω	μ	log$K_{o/w}$	A_j	B_j	α	β	AN	DN	E_b	C_b
13	14	15	16	17	18	19	20	21	22	23	24
13	12	19	22	28	28	24	23	23	18	14	14
13	12	19	22	28	28	24	23	23	18	14	14
13	12	19	22	28	28	24	23	23	18	14	14
9	9	14	17	22	22	19	19	18	17	13	13
13	12	19	22	28	28	24	23	23	18	14	14
13	12	19	22	28	28	24	23	23	18	14	14
13	12	19	22	28	28	24	23	23	18	14	14
13	12	19	22	28	28	24	23	23	18	14	14
12	10	17	18	22	22	19	19	19	13	11	11
11	12	17	21	25	25	21	20	20	14	11	11
12	10	16	21	23	23	22	21	20	15	13	13
9	10	12	14	16	16	15	15	15	12	9	9
1	8	10	9	12	12	12	12	13	10	9	9
0.960	1	9	9	12	12	10	10	11	9	7	7
−0.182	−0.095	1	15	19	19	16	15	16	12	10	10
−0.765	−0.641	0.279	1	22	22	20	19	18	13	11	11
0.903	0.913	−0.439	−0.719	1	28	23	22	22	17	13	13
0.369	0.469	0.200	−0.597	0.321	1	23	22	22	17	13	13
0.799	0.882	−0.442	−0.445	0.942	0.051	1	23	21	15	14	14
0.251	0.199	0.011	−0.758	0.295	0.361	0.240	1	20	15	14	14
0.922	0.946	−0.273	−0.574	0.980	0.248	0.953	0.239	1	16	13	13
−0.003	−0.282	0.585	−0.543	−0.054	0.403	−0.082	0.943	−0.050	1	12	12
−0.604	0.033	−0.078	−0.616	−0.129	0.405	−0.113	0.784	−0.110	0.524	1	14
−0.224	−0.559	0.584	0.232	−0.280	−0.366	−0.256	0.287	−0.427	0.406	0.041	1

contain only three or four significant terms, instead of the original number, here 24.

Parameters for which values were known for very few solvents, and solvents for which very few parameters were known, were omitted from consideration, as too many gaps vitiate the analysis. The data used nevertheless still contain many gaps. Some of these exist merely because the needed measurements have not yet been made, but in many cases the measurements are difficult or impossible, usually because of problems with solubility or chemical incompatibility. A body of data containing no gaps, when analysed by PCA, should yield a set of positive numbers, representing the decreasing importance of successive components. These are the eigenvalues of the matrix of all the correlation coefficients, that is the symmetric matrix obtained from Table 4.2 by replacing all the entries $n_{i,j}$ above the diagonal with $r_{i,j} = r_{j,i}$. A plot of these eigenvalues against the component number has a fancied resemblance to the pile of debris below a cliff, so is termed a scree plot. Such a plot for these data is shown in Fig. 4.1. Eigenvalues less than unity are commonly considered to be negligible. Here apparently the first four PCs are significant. This is perhaps unfortunate, as it is difficult to represent more than two dimensions on the page. We shall contrive to represent the first three. The scree plot also shows the consequence of the gaps in the data, in that the eigenvalues are not all positive. The last six have small negative values. These we shall ignore, though they contribute to the uncertainty.

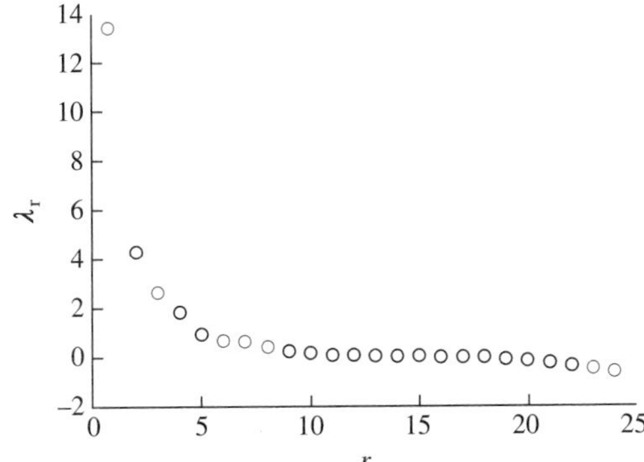

Fig. 4.1 Scree plot: eigenvalues λ_r for the correlation of 24 solvent parameters, in descending order. Four eigenvalues are greater than unity, suggesting that four independent properties of the solvent are significant.

4.3 Representation of correlation data on the hemisphere

The results of PCA of the above correlations were applied to generate a graphical representation of the similarities and differences among the parameters correlated, proceeding as follows. The first three principal components, which are mutually orthogonal by construction, are used to set up a coordinate system. Each of the original parameters p_i $(i = 1, 2, \ldots, n)$ is correlated with each of the first three PCs Z_r $(r = 1, 2, 3)$. Then each of the original parameters is considered to be a vector in the space spanned by these coordinates, with its origin at the origin of coordinates and its direction defined approximately by the set of angles it makes with the three axes. The cosine of each of these angles may be shown to be equal to the corresponding correlation coefficient. The uncertainty of direction arises from the part of the variation of the parameter associated with the fourth, fifth, etc. principal components, from experimental error, and from the gaps in the data. The magnitude of the vector does not affect the correlation coefficients, nor the angles (in fact, the magnitudes are lost in the correlation process), so the vectors can all be treated as terminating on the surface of a sphere of unit radius. In order to avoid having to show the whole surface of the sphere, it is convenient to plot as points on a hemisphere both those vectors that naturally appeared there and the 'antipodes' of those that would appear on the hidden side (if two parameters have a large correlation coefficient, they measure the same property, even though the sign may be negative).

In Fig. 4.2, a number of the parameters are plotted on a hemisphere (in projection). The first two PCs are labelled X and Y, on the equator at longitude $-45°$ and $45°$, respectively, and the third Z, at the north pole. Each point is plotted within an error circle. It may be shown that if the error is zero, the sum of the squares of the three correlation coefficients, that is, of the three direction cosines, is unity (Coxeter 1961). The error circles drawn are arbitrary, but are drawn with radii proportional to $|1 - \Sigma_{\nu=1}^{3} \cos^2\theta_\nu|$. Parameters that measure similar properties should appear close together. Measures that are essentially independent, that is to say, orthogonal, are nearly 90° apart, as the term implies. Some of the error circles are very large, implying either experimental error or substantial contributions from the fourth or later PCs.

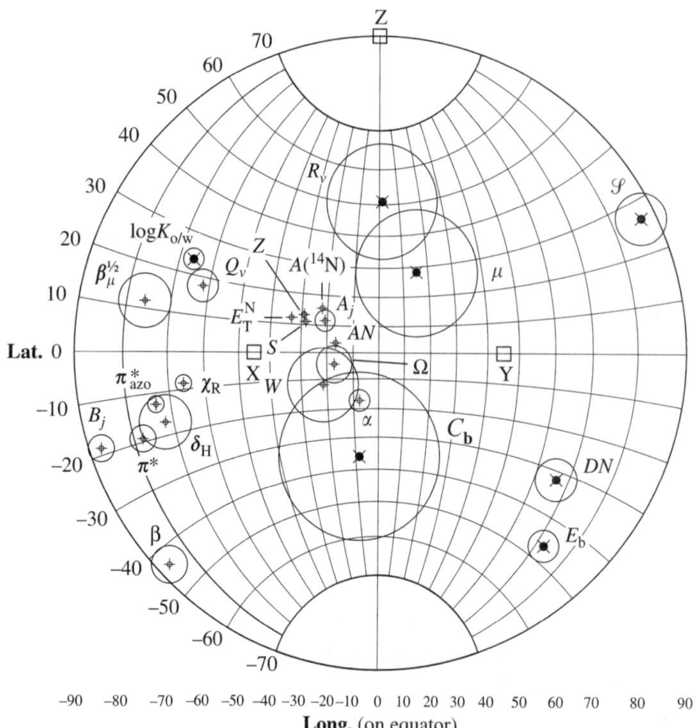

Fig. 4.2 Representation of the 24 parameters on the hemisphere. Filled circles represent positive vector directions; open circles negative, that is, antipodes. The error circles are on an arbitrary scale.

Because the first PC (X), by construction, contains the greatest part of the variance of all the data, many of the points are loosely scattered about the X axis, at $-45°$ longitude. The way the points are grouped show some unexpected similarities and differences. Notable groups are:

1. The 'acidic' group (clustered about lat. 10°, long. $-10°$): A_j, α, AN, and $A(^{14}\mathrm{N})$, with Ω and W nearby;
2. The group around lat. 20°, long. $-70°$, which contains the supposedly pure (di)polarity measures, Q_v, $\beta_\mu^{1/2}$, and $-\log_{10}K_{\mathrm{o/w}}$;
3. The basic group: C_b, DN, and E_b (all plotted as 'antipodes'), β, and B_j. All are about 90° distant from the acid group, but in different directions. The scatter of these measures about the diagram perhaps emphasizes that basicity is not a simple concept. The hard/soft classification is not sufficient to explain all the differences. The basic parameters are considered separately in Section 4.5.
4. A group around lat $-15°$, long. $-75°$, containing the solvatochromic parameters π^*, π^*_{azo}, and -χ_R, plus the heterogeneous group $\mathcal{S}$ (if it is replotted directly), Hildebrand's δ_H, and the errant basity parameter B_j. The presence of δ_H in this group may be a consequence of including the alcohols among the solvents considered, for which δ_H was assigned high values, of doubtful significance.
5. Two parameters related to polarizability, R_v (lat 40°, long. 0°) and Marcus's μ (lat 29°, long. 15°) appear fairly close together.

6. E_T^N, Z and S lie close together, between the polar and acidic groups, rather closer to the latter.

4.4 Some particular cases

Beginning in 1972, Kamlet, Taft and coworkers carried out a long series of studies[2] of solvent effects, beginning with the effects on spectra (frequency shifts and intensities). They show that it is necessary to consider not only the bulk polarity of the solvent, but also the abilities of the solvent to act as donor or acceptor in hydrogen bonding. They use a parameter π^* to represent the bulk polarity, choosing this symbol because it best correlates frequency shifts of $\pi \rightarrow \pi^*$ and $n \rightarrow \pi^*$ electronic transitions. The hydrogen-bonding ability is represented by two parameters: α for the solvent's tendency to act as a proton donor (acidic character), and β for its proton acceptor tendency (basic character). They then write eqn 4.4 (cf. eqn 4.1):

$$XYZ = XYZ_0 + s\pi^* + a\alpha + b\beta \tag{4.4}$$

where the symbol XYZ represents the property being studied (expressed in such a way that it is proportional to an energy, for example, a spectral frequency or the logarithm of a rate or equilibrium constant), the subscript 0 indicates its value in the reference solvent, π^*, α, and β are the solvent parameters defined above, and s, a, and b are the sensitivity coefficients for the property XYZ to the corresponding solvent property. Values of π^*, α, and β for some solvents are listed in Appendix Table A.3.

For an example of a pairwise correlation, consider the following two quantities that purport to measure polarity of the solvent. Kosower (1958, 1968) proposed the parameter Z, based on the change of frequency of a charge-transfer transition of 1-ethyl-4-methoxycarbonyl-pyridinium iodide (**1**), which shows a pronounced increase of frequency (hypsochromism) when the polarity of the solvent is increased, because the ionic ground state is stabilized by a polar medium relative to the non-ionic excited state. Dimroth and Reichardt (1969; Reichardt and Harbusch-Görnert 1983) proposed another, $E_T(30)$, or in 'normalised' form, E_T^N, based on a visible transition of a pyridinium-*N*-phenoxide betaine dye (**8**, p. 40) (the label (30) is because the dye was the 30th of a number tried). Figure 4.3 shows the correlation of these two parameters for 25 solvents.

O OCH$_3$ +N I$^-$

1

The correlation does not appear to be very good, but in fact, by using a select set of solvents believed to be free from interferences such as specific interactions or purity problems, it has been used to obtain estimates of $E_T(30)$ for solvents in which the dye is insoluble or reacts. With these additions, $E_T(30)$ was recently available for 271 solvents (Reichardt 1994).

[2] For example, Kamlet and Taft (1979); Kamlet *et al.* (1979); Abboud and Taft (1979); Kamlet *et al.* (1981); Taft *et al.* (1981).

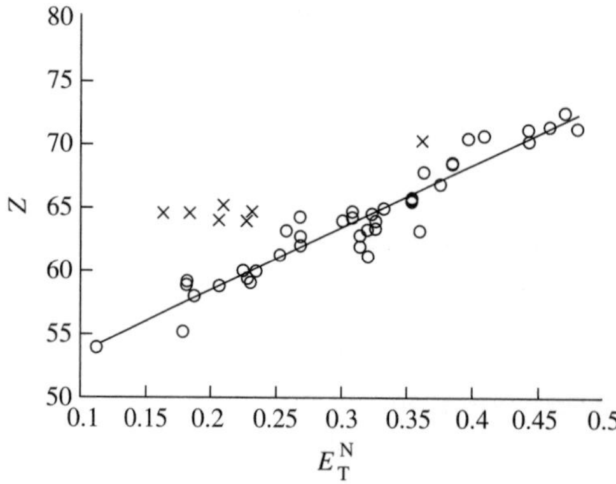

Fig. 4.3 Kosower's Z versus Reichardt's E_T^N. Data from the compilation of Abboud and Notario (2000). Points marked by crosses are omitted from the statistics. They are for compounds flagged by those authors as uncertain, plus a group of ethers and esters that seem to have values of Z clustered about 64–65, lying above the trend. No strong HBD solvents are included.

Another example with a different result was the attempted correlation of the theoretically-justified quantity β_μ, defined by Dutkiewicz (1990) as μ^2/V (the square of the dipole moment of the solvent molecule divided by the molar volume), with E_T^N. She found that, instead of lying close to a single line, the plotted points fell into four fairly distinct classes (with a few outliers). The main classes were:

1. Weakly dipolar and non-polar, non-HBD (hydrogen-bond donor) molecules: hydrocarbons and halogenated derivatives, ethers, esters, tertiary amines.
2. Strongly dipolar, non-HBD: ketones, *N,N*-dialkylamides, nitrocompounds, sulphoxides, sulphones, pyridine.
3. HBD molecules: water, primary alcohols, glycols, carboxylic acids, with non-primary alcohols and aniline and perhaps chloroform (weak HBDs) forming subclass 3a, and phenols (strong HBDs), 3b.
4. *N*-monoalkylamides and formamide (strongly dipolar, weak HBDs).

Figure 4.4(a) is similar to Dutkiewicz's figure, but with the axes exchanged.

The distinction among these classes seems to rest mainly upon their ability to function as donors in the formation of hydrogen bonds (HBD). To test this, the current authors have tried a bilinear correlation of E_T^N with β_μ and the parameter α, derived by Kamlet and Taft (discussed above), which is taken as a measure of HBD strength. The results may be expressed by eqn 4.5, with the constants $C_0 = 0.1358$, $C_1 = 0.004085$, and $C_2 = 0.5022$; the standard error of fit was 0.056.

$$E_T^N = C_0 + C_1\beta_\mu + C_2\alpha \qquad (4.5)$$

Figure 4.5 shows a plot of E_T^N versus the composite variable, $(C_1\beta_\mu + C_2\alpha)$.

A plot of $(E_T^N - C_2\alpha)$ versus β_μ, expected to be linear, showed some curvature; we therefore tried $\sqrt{\beta_\mu}$. Figure 4.6 shows the result of plotting E_T^N versus the new composite variable, $(C_1'\beta_\mu^{1/2} + C_2'\alpha)$, with $C_1' = 0.04324$ and

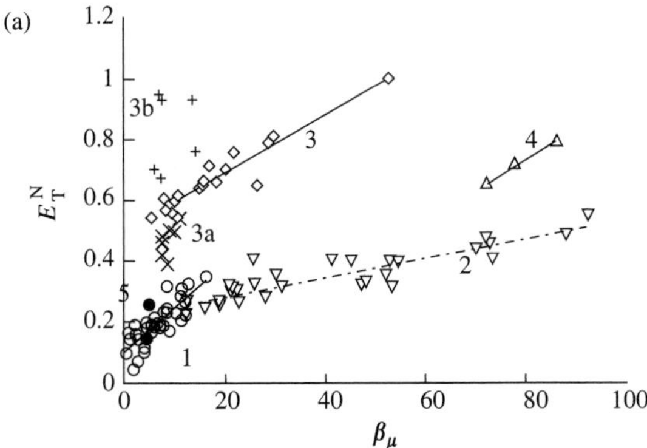

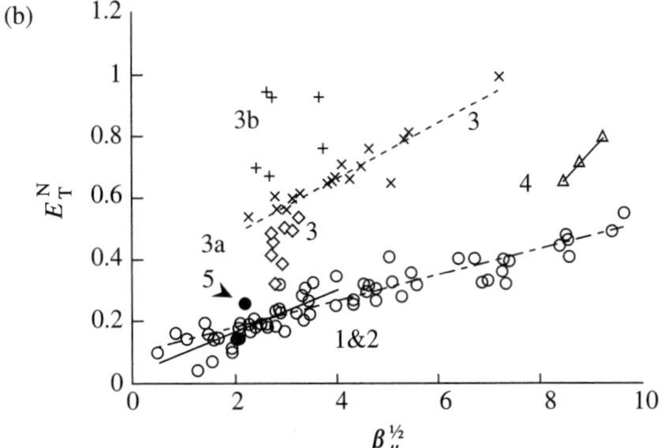

Fig. 4.4 (a) E_T^N versus. β_μ Classes indicated by numbers: (1), etc. Redrawn after Dutkiewicz (1990), with permission. The axes have been interchanged. (b) E_T^N versus the square root of β_μ. Classes 1 and 2 are treated as one class.

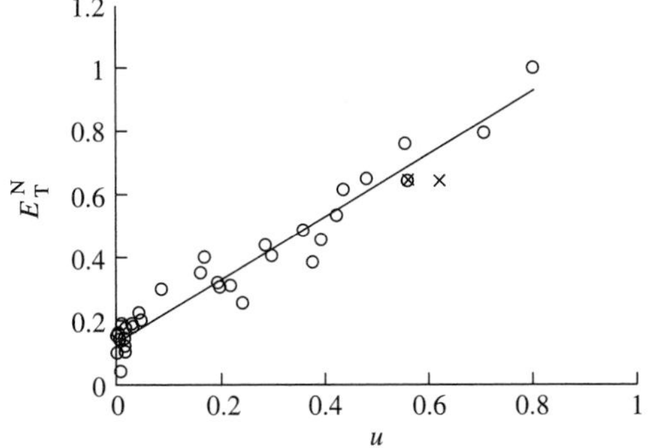

Fig. 4.5 E_T^N versus the composite variable: $u = C_1\beta_\mu + C_2\alpha$; $C_1 = 0.04085$, $C_2 = 0.5022$. In this and the following figure, the points marked by crosses are for acetic acid. The circled cross is for the dimer; the plain cross is for the monomer, not included in the statistics.

$C_2' = 0.4780$. ($C_0' = 0.06112$). The fit is not notably better judged by the standard error, now 0.052, chiefly owing to the general scatter, but the linearity is improved, especially near the origin. One may suggest (without theoretical justification) that the square root of β_μ is a better measure of the contribution of the dipole moment to the overall 'polarity' than β_μ itself. Since α-values are not available for all the solvents in Dutkiewicz's classification, there are fewer points plotted, but it appears in Fig. 4.4(b), in which $\sqrt{\beta_\mu}$ replaces β_μ, that classes 1 and 2 are no longer clearly separated, but are

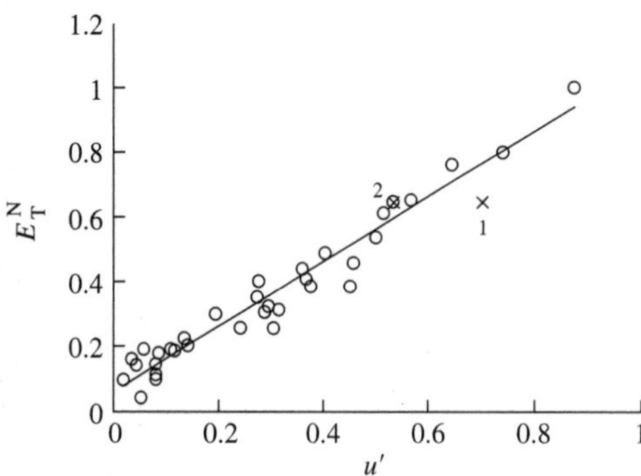

Fig. 4.6 E_T^N versus the composite variable: $u' = C_1'\beta_\mu^{1/2} + C_2'\alpha$; $C_1' = 0.0432$, $C_2' = 0.4780$. See caption for Fig. 4.5.

scattered about one line. Since they are similar in lacking HBD ability, and differ chiefly in dipolarity, this is not surprising.

Carboxylic acids present a problem in assigning β_μ. Acetic acid, for instance, associates strongly through hydrogen bonding to form the symmetrical dimer (**2**) in the pure liquid or in solution in non-hydroxylic solvents. Should the value of the dipole moment, μ, be that of the monomer (1.74 Debye, or 5.6×10^{-30} C m, in the gas phase) or that of the dimer, which is zero? Perhaps it should be assigned some intermediate, probably variable, value representing the effect of the few free monomers or the tendency of the dimer to dissociate in the vicinity of a polar solute molecule with which it can interact. In each of the Figs 4.5 and 4.6, acetic acid appears twice, marked by crosses: once with $\beta_\mu = 14.97$ kJ mol^{-1} (plain cross), once with zero (circled cross in the figures). In both Figs 4.5 and 4.6, it is the point (2) for the dimer that seems closest to the trend of the other points.

O----H–O
CH_3–C   C–CH_3
O–H---O

2

Swain *et al.* (1983) have proposed a pair of parameters, A_j and B_j, called *acity* and *basity*, coinages to imply similarity to acidity and basicity, but properties of the bulk solvents rather than the molecules. They fitted data for a great many phenomena (reaction rates, equilibria, spectral shifts) in a variety of solvents to equations of the form of eqn 4.6.

$$p_{i,j} = c_i + a_i A_j + b_i B_j \tag{4.6}$$

where $p_{i,j}$ and c_i have the dimensions of energy. Four fixed points served to anchor the scales. These were: for water, $A_j = B_j = 1$; for n-heptane, $A_j = B_j = 0$, and two additional conditions: for hexamethylphosphortriamide (HMPA) $A_j = 0$; for trifluoroacetic acid $B_j = 0$.

Acity represents some composite of all the properties of a substance that are associated with positive charge: HBD activity (tendency to donate a shared proton in a hydrogen bond), Brønsted acidity (tendency to donate a proton outright), Lewis acidity (tendency to receive a shared electron pair), electron affinity (tendency to receive an electron outright), oxidizing strength, electrophilicity; in a phrase, Usanovich acidity. Basity similarly represents a composite of all the properties that are associated with negative charge: Brønsted or Lewis basicity (tendency to donate a shared electron pair to a

Table 4.3 (a) Coefficients of determination* for partial dependence of E_T^N on. α and either β_μ or $\beta_\mu^{1/2}$

Dependent variables	$R^2_{\beta\mu}$ **	R^2_α
β_μ, α	0.802	0.921
$\beta_\mu^{1/2}$, α	0.826	0.928

* For data fitted by $y = a + bx$, $R^2 = (a\Sigma y_i + b\Sigma x_i y_i - (\Sigma y_i)^2/n)/(\Sigma y_i^2 - (\Sigma y_i)^2/n)$.
** To the appropriate power.

Table 4.3 (b) Coefficients of correlation between E_T^N (1) and A_j (3), B_j (4), and either β_μ or $\beta_\mu^{1/2}$ (2)

Dependent variables	$R^2_{1,2}$ *	$R^2_{1,.3}$	$R^2_{1,4}$
β_μ, A_j & B_j	0.544	0.940	0.178
$\beta_\mu^{1/2}$, A_j & B_j	0.731	0.947	0.039

* To the appropriate power.

proton or to a vacant orbital, such as on a metal), the negative of ionization potential (tendency to lose an electron outright), reducing strength, nucleophilicity; in sum, Usanovich basicity.

We have correlated E_T^N simultaneously with A_j, B_j, and with β_μ both directly and as its square root. Table 4.3 shows the coefficients of determination for these data. Again, the square root of β_μ gives an improved correlation, and acetic acid fits the trend better if assigned $\beta_\mu =$ zero. It appears that in bulk acetic acid the dimer is stable enough to cause the liquid to act as a non-dipolar medium, but labile enough to allow strong interaction with other, H-bond acceptor (HBA) molecules. It may be noted that when the square root of β_μ is used, the correlation coefficient with B_j becomes 0.039, which is effectively zero. In all the foregoing sets of correlations, E_T^N turns out to be as closely related to the acidity (in the guise of HBD strength or of acity) as to the (di-)polarity. It is essentially uncorrelated with polarizability, as measured by the polarization function: $R_v = (n^2 - 1)/(n^2 + 2)$.

4.5 Acidity and basicity parameters

The two ac(id)ity parameters α and A_j are strongly correlated ($r = 0.974$ using a set of fourteen solvents for which both quantities are non-zero). Within the set, though, there seems to be a difference in behaviour between C-acids (chlorohydrocarbons, ketones, nitromethane and acetonitrile) on the one hand and O-acids (five alcohols and water) on the other. The slope obtained when values of α for the C-acids are plotted versus A_j (Fig. 4.7) is about 1.5 and the intercept negative, whereas the O-acids gave a slope near unity and an intercept (within the rather large uncertainty) zero. A_j values are known for a number of solvents to which zero values of α have been assigned. Extrapolation of the regression line (using O-acids only) suggests negative values of α for these solvents. Since negative acidity in this context appears meaningless, it may be that a shift of the zero of the α scale is in order. In any case, α and A_j appear to measure somewhat different aspects of acidity, as their authors intended.

Catalán (2001) records values of three parameters for 190 solvents. They are *SPP*, believed to be a measure of pure polarity uncontaminated by acidity, *SB*, a measure of basicity, and *SA*, a measure of acidity. *SPP* is based on the differences $\Delta\nu$ between the frequency of the first absorption maximum between 2-(*N*,*N*-dimethylamino)-7-nitrofluorene (DMANF, **3**) and 2-fluoro-7-nitrofluorene (FNF, **4**), according to eqn 4.7.

$$SPP(\text{solvent}) = \frac{\Delta\nu(\text{solvent}) - \Delta\nu(\text{gas})}{\Delta\nu(\text{DMSO}) - \Delta\nu(\text{gas})} \tag{4.7}$$

SB is similarly based on the difference in frequency between 5-nitroindoline (NI, **5**) and 1-methyl-5-nitroindoline (MNI, **6**), according to eqn 4.8 (TMG is tetramethylguanidine).

$$SB(\text{solvent}) = \frac{\Delta\nu(\text{solvent}) - \Delta\nu(\text{gas})}{\Delta\nu(\text{TMG}) - \Delta\nu(\text{gas})} \tag{4.8}$$

The acid parameter, SA, is based first on the relation between the observed absorbance frequency maxima, measured in non-acidic solvents, of the stilbazolium betaine dye **7** (TBSB: R = H) and the O,O'-di-*t*-butyl derivative

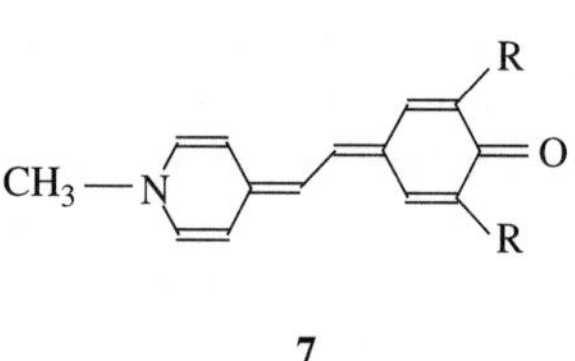

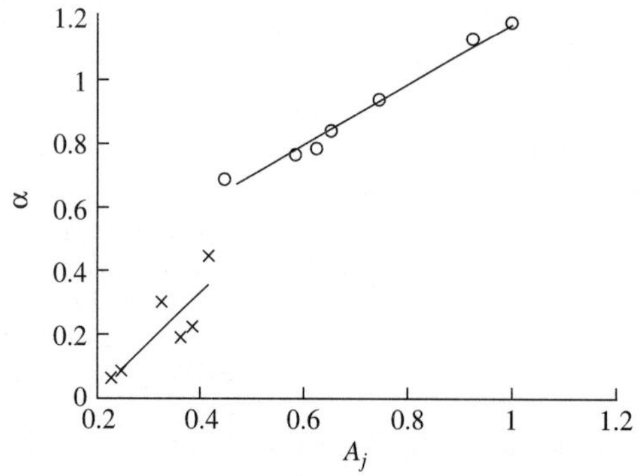

Fig. 4.7 Acidity parameters: Taft–Kamlet α versus Swain's A_j. Circles are for O-acids, crosses for C-acids. Separate least-squares lines are fitted; see text. (Data from Kamlet *et al.* 1983 and Swain *et al.* 1983.)

(DTBSB: R = *t*-butyl), represented by eqn 4.9.

$$\nu_{TBSB} = 1.409\nu_{DTBSB} - 6288.7 \qquad (4.9)$$

Then, for acid solvents,

$$SA = 0.400\frac{(\nu_{TBSB} - (1.409\nu_{DTBSB} - 6288.7))}{1299.8} \qquad (4.10)$$

which fixes the value for ethanol at 0.400. *SA* shares with α the assignment of zero to a number of solvents that have non-zero values of A_j. A_j and *SA* are compared in Fig. 4.8, plotting only those points for which both measures are non-zero. There is a reasonable correlation between them for the O-acids, but the remainder of the points cluster about a line with nearly infinite slope. A similar plot of α against *SA* gave similar results, the alcohols following a linear trend of slope 0.55 (with the exception of 2,2,2-trifluoroethanol) and the C-acids in a loose cluster.

As noted above, there are also some differences among the various basic parameters. The values of the basic parameters β and B_j (considering only solvents having non-zero values of both) are almost totally uncorrelated ($r = 0.148$, 17 solvents); they seem to measure quite different aspects of basicity. In addition to these two, there are Drago's E_b and C_b (Drago and Wayland 1965), which purport to measure covalent and electrovalent contributions to base strength, Gutmann's (1967) donor number *DN*, the soft base parameter D_S (Persson *et al.* 1987), Maria and Gal's (1985) $\Delta H^0(BF_3)$, and Catalán's (2001) *SB*, eight in all. We have carried out a PCA analysis of these eight parameters, using data for 19 solvents for which most are known. The first three eigenvalues of the correlation matrix, in decreasing order, were 4.52, 1.80, and 1.01, with the rest less than 0.5, suggesting that there are three distinct components, the first being dominant, as it represents more than half the total variance of the data (4.52/8).

Figure 4.9 shows the eight base parameters on a hemisphere, with the three PCs as axes. Four of them (C_b, *DN*, D_S, and $\Delta H(BF_3)$) form a loose cluster.

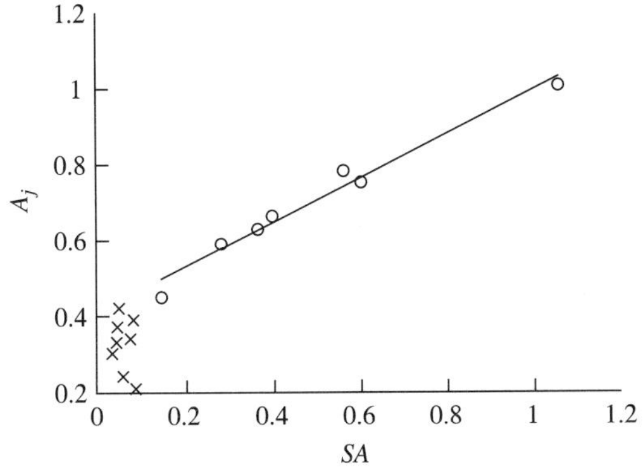

Fig. 4.8 Swain's acity parameter A_j versus Catalán's (2001) *SA*. Circles are for O-acids (alcohols); crosses for assorted C-acids. The plotted line, fitted to the O-acids only, is represented by:
$A_j = 0.414 + 0.573\ SA$; $S = 0.032$

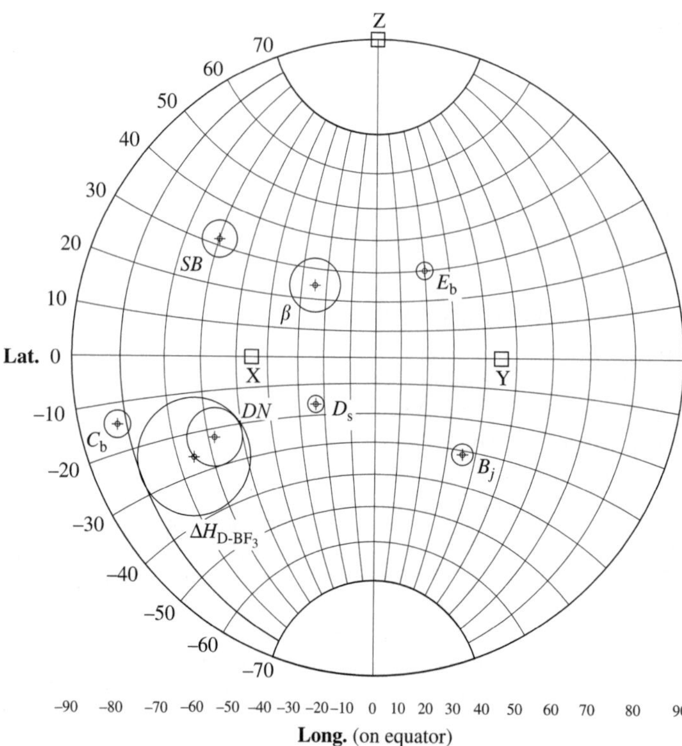

Fig. 4.9 Eight basic parameters. Parameters represented are: (latitude, longitude in degrees)

1. β (Taft *et al.* 1985; 25°, −25°).
2. B_j (Swain *et al.* 1983; −31°, 35°).
3. *SB* (Catalán 2001; 32°, −60°).
4. *DN* (Gutmann 1967; −22°, −59°).
5. C_b (Drago 1980; −14°, −82°).
6. E_b (Drago 1980; 30°, 20°).
7. δH_{D-BF_3} (Maria and Gal 1985; −25°, −66°).
8. D_s (Persson *et al.* 1987; −16°, −24°).

C_b, since it purports to measure the covalent part of the acid/base interaction, and D_S, which is explicitly designed to measure soft basicity, clearly belong together. Why *DN* and ΔH(D-BF_3) are in this group is less clear, as SbF_5 and BF_3 are not soft acids. *SB*, β, and E_b form another loose cluster, centred some 70° from the first. They all represent measures of hard basicity: *SB* is based on the interaction of the base with the N-H acid, 5-nitroindolene, a hard acid; β measures HBA activity, so is hard, and E_b was designed to measure the electrostatic (hard) contribution to basicity. The basity parameter, B_j, lies apart from all the rest, as was noted, in Section 4.3. This analysis reinforces the idea that 'basicity' is not a simple property.

4.6 Softness parameters

Chen, *et al.* (2000) consider the relationships among three solvent softness scales relevant to bases, namely, Marcus's μ (discussed in Chapter 3), D_S (above), and a third measure, symbolized by $\Delta\Delta\nu$(I-C), which is the difference between the shifts of the C-I stretch in the molecule ICN and 0.085 times the shift of the O–H stretch in phenol, in the same solvents. To these we add the volume polarization R_v (for softness is conceptually related to polarizability), $D = 0.086C_b - 0.76E_b$, suggested by Fig. 3.3b, and S_{orb}, the reciprocal of the LUMO–HOMO energy difference, suggested by Klopman (Klopman 1968; Huheey *et al.* 1993: 351) as further measures of softness.

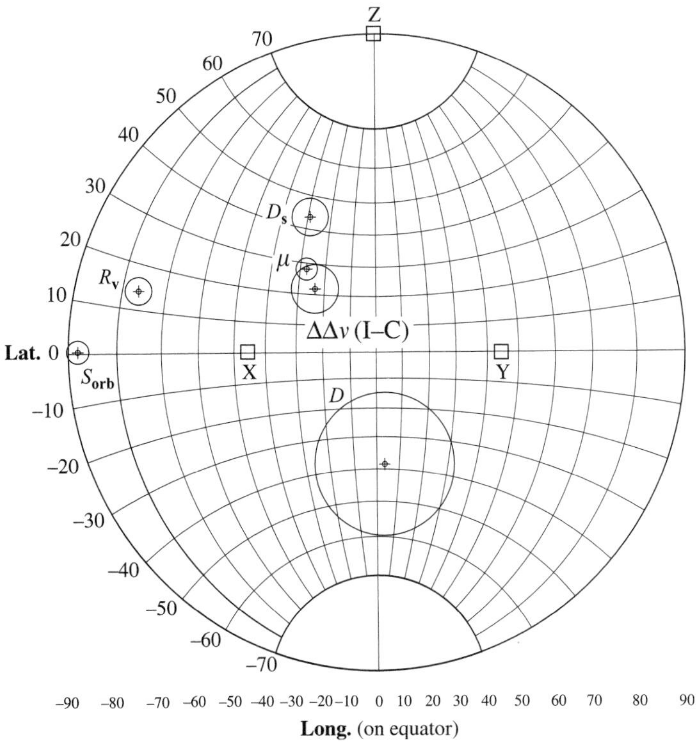

Fig. 4.10 Six softness parameters. Three (μ, D_S, and $\Delta\Delta\nu$(C − I) are discussed by Chen *et al.* 2000). R_v is the volume polarization, D is derived from the Drago parameters C_b and E_b, and S_{orb} is the reciprocal of the LUMO–HOMO energy difference (values from Mu *et al.* 1998).

Figure 4.10 shows these six on a hemisphere projection, using a new, appropriate set of coordinates. When they were subjected to PCA, the scree plot showed one component as representing the great part of the variance, with two very minor components and some 'noise'. As expected, the six vectors are reasonably well clustered about the first PC, *X*. $\Delta\Delta\nu$(I-C),μ, and D_s are more closely associated at one extreme, and R_v and S_{orb} at the other. The latter pair are properties of the whole molecule, as noted above, in contrast to the three others, which are properties of the basic site on the molecule. *D* was available for only seven of the 35 solvents in the set, so it is not surprising that it is subject to large uncertainty, and lies apart from the rest.

4.7 Conclusion

The statistical methods of chemometrics, particularly Principal Component Analysis, have been helpful in sorting out confusing data in many aspects of chemistry, and linear free energy relationships in particular have been fruitful in providing insight into reaction mechanisms. The calculations discussed in this chapter only begin to show what could be done to aid understanding of some of the empirical parameters applied to the study of solvent effects. These techniques afford views of reactions that resemble impressionist paintings. For full understanding of the way in which reactions occur, whether in the gas phase or in solution, we must wait until theories of elementary reactions (reaction dynamics) and of the structure and energetics of solvents and solvates have reached suitable levels. Progress in these matters is now rapid. In the next chapter we hope to show the direction in which the theory of solutions is moving, and to provide a brief introduction to the ways in which some theoretical techniques can be applied to problems in the field of reactions in solution.

5 Theoretical calculations

5.1 Introduction: Modelling

The purpose of computer-based modelling of chemical systems is to mimic them in properties of interest, and to understand at a molecular or even electronic level the physical origins of these properties. The ultimate goal in modelling a reaction in solution is to show, by modelling, the interaction of dissolved reactants, activated complexes and products with solvent molecules, how changing the solvent affects the rate or equilibrium of a reaction. In some cases we may learn why a solvent may favour one of a set of possible mechanisms. To study the solvation of a single solute species, we may consider a two-stage model representing, first, an assembly of solvent molecules plus an isolated solute molecule or ion and, second, the system consisting of the same solute surrounded by the solvent. When suitably translated into molar quantities, the changes in U, H, S, G, and γ then yield estimates of the transfer energy, enthalpy, entropy, Gibbs energy and activity coefficients of the solute from the ideal gas phase to the dilute solution. Though in practice this goal is not generally attainable, the calculations often provide valuable insight.

Of the available methods, *quantum mechanics* (QM) attacks the problem at its deepest level. Moore (1972), in one edition of his physical chemistry text, says that, in principle, all of chemistry could be calculated from the Schroedinger equation. Then in a footnote, he adds: ' "In principle" from the French, "*En principe, oui*", which means, "*Non*".' Since that date, however, computers and programs have become more powerful, and much effort is being made to carry out quantum-mechanical calculations of the energetics of solvation of molecules and ions in various solvents. QM calculations are implemented in *ab initio* form at various levels of approximation, semi-empirically also at various levels, and, more recently through density-functional theory.

Other methods treat the molecules as classical objects, interacting through forces of various kinds, and use the formalism of statistical thermodynamics to obtain the desired quantities as averages. The chief of these methods are *Monte Carlo* (MC), so called because it depends on (simulated) chance in the form of random numbers in evaluating average properties, and *Molecular Dynamics* (MD), which uses the laws of motion to explicitly represent the evolution in time of an assembly of molecules giving the desired properties as time averages. There are also various hybrid methods, in which, for instance, the translation and rotation of molecules are assumed to behave classically, but the internal vibrations are treated quantum-mechanically, or the solute is treated quantum-mechanically but the solvent molecules treated classically, and methods in which solute molecules are treated as neither *in vacuo* nor surrounded by other molecules, but as in a cavity in a continuous dielectric.

Finally there is a group of methods labelled *integral equation theories*. These begin with the pairwise potential energy function, which may be derived from quantum calculation, or may take one of several forms that are more or less realistic. The simplest is the 'hard sphere' potential, which is zero for separations greater than an assumed collision distance, and infinite within that distance. Other forms of potential are mentioned in Section 5.3 (see Berry *et al.* 2000: 305). The radial distribution function, $g(\boldsymbol{r})$, measures the departure from the mean of the number density of molecules around one molecule taken as fixed at the origin, as a function of the radius vector $\boldsymbol{r}$. The integral equation theories are so called because $g(\boldsymbol{r})$ is obtained from coupled equations involving integrals.

We shall not attempt to describe any of these methods in detail, but offer outlines of them, and take a look at some particular results.

5.2 Quantum-mechanical methods

5.2.1 *Ab initio*

Quantum mechanical calculations on single molecules can be done at different levels of theory. The most fundamental approach, the so-called '*ab-initio*' ('from the beginning') method (Hehre *et al.* 1986), begins with a guess at the structure of a molecule, guided by experience. The *Schroedinger Equation* can be solved exactly for hydrogen-like atoms (a nucleus and one electron only). It can be solved nearly exactly for the hydrogen molecule-ion H_2^+, subject only to the *Born–Oppenheimer approximation*: since nuclear motions are three orders of magnitude slower than electronic motions, the positions of the nuclei are assumed to be fixed, and the electrons treated as moving in the field created by the fixed nuclei. For other, not-too-large atoms and for small molecules the SE can be solved to arbitrary closeness of approximation, with due effort. For larger systems, the effort, that is to say the requirement of computer time and memory, grows rapidly with increasing numbers of electrons and nuclei, so lower levels of approximation become necessary. This is generally true whether the system consists of a single large molecule or more than a very few smaller molecules in interaction, though certain methods have been applied successfully to large systems.

The Born–Oppenheimer approximation is universally applied in calculations relevant to solvation. Electrons are accommodated in molecular orbitals, which are represented as *linear combinations of atomic orbitals* (LCAO). The atomic orbitals chosen form the *basis set*. One then calculates the energy of the molecule (or collection of molecules), using the variation principle (see below) to obtain the lowest ('best') energy for the chosen basis set and molecular geometry. Repetition of the whole calculation with slightly shifted nuclei enables one to find the configuration of minimum energy and the shape of the hypersurface representing the energy as a function of nuclear coordinates in the vicinity of the minimum. As far as nuclear motions are concerned, this hypersurface represents what is effectively potential energy. From it the vibrational frequencies and the *zero-point energies* associated with them can be obtained. The energy of the molecule at 0 K is the energy minimum plus the zero-point vibrational energies.

In the simplest method, the *Hartree–Fock* (HF) or *self-consistent-field* (SCF) method, each electron is treated as if it moves in an average electric field due to the nuclei and all the other electrons. The calculations may be done, in the simplest version, by taking as the basis set just enough atomic orbitals to represent the core and valence shells of each atom: for example, for the water molecule, the 1s, 2s, $2p_x$, $2p_y$, and $2p_z$ on oxygen and a 1s on each hydrogen, seven in all, which can be combined to form seven molecular orbitals. This is the *minimum basis set*. The orbitals themselves may be of the *Slater type*, which are simplified versions of hydrogen-like orbitals, with the oscillations in the radial part of the functions smoothed out. In current practice they are almost always of the *Gaussian type*, which were introduced as approximations of Slater or hydrogen-like orbitals by linear combinations of several Gaussians (see Fig. 5.1), which greatly simplifies the calculations (chiefly because the product of two Gaussians is another Gaussian: Fig. 5.2).

Gaussian approximations necessarily have two failings: they underestimate the magnitude of the wave function both at $r = 0$ (underestimating both charge and spin density at the nucleus) and at large r (underestimating interatomic interactions through overlap). Fewer than three Gaussians per orbital make these errors unacceptably large. The simplest minimum basis set that is at all usable is designated STO-3G (Slater-Type Orbitals –3 Gaussians).

The minimum basis set may be augmented by the addition of further orbitals, the purpose of which is to correct the shapes of the orbitals

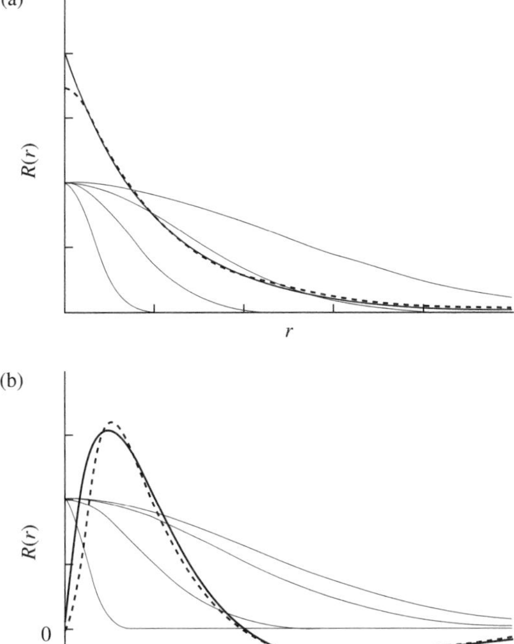

Fig. 5.1 Illustration of the approximation of radial parts of hydrogen-like atomic orbitals as sums of Gaussians: (a) 1s; (b) 3p. Not to scale, and fitted by trial and error (not optimized). The light curves are the Gaussians, the dotted curve their weighted sum. Note that in each case the magnitude of the orbital is underestimated near the nucleus.

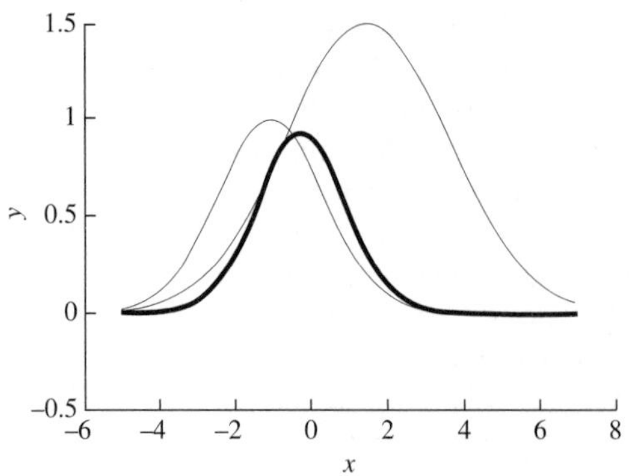

Fig. 5.2 Two Gaussian functions in one dimension, G_1 and G_2, and their product (heavy curve).

especially as they may be distorted in the molecular environment, to make them more realistic. Designations such as 6-31G* describe some of these *extended basis sets*. The reader may consult Hehre *et al.* (1986), Grant and Richards (1995), Leach (1996), Levine (2000), or the review by Davidson and Feller (1986) for specific information.

The *Variation Principle* ensures that the more adjustable parameters the wave function contains, of course within the constraints of the problem, the lower the calculated minimum energy will be, and that this minimum will not be lower than the true energy. How far one pushes the elaboration of the assumed form of the wave function by adding other elementary functions to the minimum basis set depends on how accurate a result is needed, but it is severely limited by the number of electrons in the molecular system under investigation. The time to complete the calculation goes up rapidly as the number of electrons and the size of the basis set are increased, so rapidly that the calculation can become impossible for a molecule that in other contexts is considered of only moderate size. This remains true even though the symmetry of the molecule is used to eliminate duplicate calculations of integrals that must be equal and to identify any that are necessarily zero, by the methods of group theory. Similar problems obviously exist in systems containing more than a very few ordinary-sized molecules.

Increasing the size of the basis set still leaves the electrons moving in an average field, and allows the calculation to approach, not the true energy, but the (higher) *HF limit*. Electrons avoid each other for two reasons. First, the *Pauli Exclusion Principle* forbids two electrons of the same spin to occupy the same spatial orbital. This is ensured if the overall wave function is antisymmetric, in the sense that exchange of any two electrons changes the sign of the wave function. Equation 5.1 is a simple example:

$$\Psi = \psi_1(1)\psi_2(2) - \psi_1(2)\psi_2(1) = \begin{vmatrix} \psi_1(1) & \psi_1(2) \\ \psi_2(1) & \psi_2(2) \end{vmatrix} \tag{5.1}$$

The two orbitals ψ_1 and ψ_2 may be different space functions or identical space functions but with opposite spin. Each may be a function of the coordinates of either electron 1 or electron 2. Interchanging the electrons changes the sign only of Ψ. For larger systems, writing the wave function in determinant form exploits two properties of any determinant: that exchanging any two columns changes its sign, and that if two columns are identical, the determinant vanishes, so that if two electrons of the same spin occupy the same region of space, the wave function will vanish. The HF method takes this constraint into account.

Secondly, Coulomb repulsion causes electrons of whatever spin to keep apart. The HF method does not properly allow for this second effect. There are several ways to do this, of which two that improve upon the HF result are in common use. *Configuration interaction* (CI) involves the inclusion of one or more electron configurations in which an electron is placed in an otherwise empty orbital (an excited state). It accounts better for electron–electron repulsion, that is, for the fact that each electron sees the others as moving particles, rather than an average cloud of charge. Full configuration interaction is feasible only for small molecules, and even then, only when modest basis sets are used. Various restricted versions are often used. Møller and Plesset (1934) proposed a method based on perturbation theory, which does not increase in complexity for larger systems (e.g. see Leach 1996: 83–85). Versions of the *Møller–Plesset* method are designated MP2, MP3, etc., indicating the order of perturbation used (first-order perturbation merely reproduces the HF energy). Compared to CI, it has the disadvantage that it does not conform to the variational principle, that is to say the lowest energy calculated may be below the true energy, though this can be compensated for. It has the advantage of much greater speed. A third method of accounting for inter-electron repulsion is density functional theory, discussed below in section 5.2.3.

If the molecule under study is polyatomic, the whole process may then be repeated with slightly different internuclear distances and angles, until a configuration of minimum total energy is found. Care must be taken to ensure that this minimum is the *global minimum*, not a local minimum that represents an metastable configuration of the molecule or an isomer. If the 'molecule' is a transition state, the point sought is a minimum with respect to all but one of the vibrational modes. It represents a maximum in the energy along the reaction coordinate, the x direction in Fig. 1.4. Vibrational analysis can tell one whether a stationary point is a minimum (all force constants positive), a transition state (one negative force constant), or a point where more than one force constant is negative.

To represent a reaction it is necessary to carry out these calculations for the reactant species and for the products. The energy change for the reaction, ΔU, is obtained as a rather small difference between large numbers, so the requirements of accuracy in the calculations are severe.

Investigation of kinetic parameters of a reaction is also possible, either by postulating a reasonable structure for the transition state, or by exploring paths of lowest energy to find saddle points (see Section 1.7). The activation energy is obtained by difference between the energy of the saddle point and that of the initial state, with corrections for the zero-point energies, as before. As in calculation of ΔU of reaction, for $\Delta U^{\ddagger}$ high accuracy is required.

Methods have been devised that combine low-level treatments of the whole molecules involved in the reaction with high-level treatments of 'model systems' containing the portions of the molecules directly involved (Levine 2000: 617–618).

5.2.2 Semi-empirical methods

To reduce the amount of calculation required, a variety of *semi-empirical* methods have been proposed. One may assume, for instance, that the core electrons (all but the valence electrons) do not change significantly in energy or in distribution about the nuclei to which they are most strongly bound, when the atoms or groups of atoms containing them are transferred from one molecule to another of related structure. It is thus possible to reduce the number of electrons for which explicit quantum-mechanical calculations are necessary, with great savings in computation time. The atomic cores are represented by fixed charge distributions, adjusted to give correct results in simple systems. Further savings result from replacing as many integrals as possible by fixed numbers, similarly adjusted. Overlap integrals between wave functions of certain symmetries on the same atom may be taken as constant; many are zero. Overlap integrals between orbitals on adjacent atoms in a molecule are assumed to depend only on the kinds of atoms involved and their separation, and overlap integrals between non-adjacent atoms may be completely neglected. The more integrals in the *ab-initio* scheme can be replaced with empirical parameters or set equal to zero, the greater the saving of computer time. Particular methods are labelled as, for example, MNDO (*modified neglect of differential overlap*) (Dewar and Thiel 1977). It is common to use semi-empirical methods to find a configuration of minimum energy, which may then be refined by *ab initio* calculation.

5.2.3 Density-functional theory

The Hohenberg and Kohn Theorem (Hohenberg and Kohn 1964; Koch and Holthausen 2001; Leach 1996: 528–533; Levine 2000: 573–592) states that all the properties of a molecular system in its ground state can be derived from the electron-density distribution function. The total energy may be expressed as the sum of kinetic, potential and exchange/correlation terms as in eqn 5.2, where ρ is to be understood as a function of the internal coordinates, symbolized by the vector $\boldsymbol{r}$.

$$E(\rho) = T(\rho) + E_{\mathrm{NN}}(\rho) + E_{\mathrm{Ne}}(\rho) + E_{\mathrm{ee}}(\rho) + E_{\mathrm{XC}}(\rho) \tag{5.2}$$

The terms, respectively, are the electronic kinetic energy, the internuclear repulsive energy, the nuclear–electronic attractive energy, the interelectronic repulsive energy, and the exchange and correlation energy. The second, third, and fourth terms are easily seen to be functionals of electron density distribution only (a function turns one number into another number: a *functional* turns a function into a number; see Becke 2000, Lewars 2003). The Hohenberg–Kohn Theorem gives assurance that calculation of the other terms is possible (but does not prescribe how). The total energy so derived is variational, that is to say, it conforms to the *Variational Principle*: an incorrect density distribution yields an energy above the true value provided the exact functional, which gives the relationship between ρ and T, etc., is known.

An approximation of the electron density may first be derived by a simplified *ab initio* calculation, in which the electrons are treated as not interacting. The electron density is obtained from the sum of the density contributions from each of the n occupied spin-orbitals, that is:

$$\rho(\boldsymbol{r}) = \sum_{i=1}^{n} (|\psi_i(\boldsymbol{r})|^2) \tag{5.3}$$

Refinement of the density function, and consequently of the energy and other related properties, is then performed, using a representation of the electron density through a set of orthonormal single-electron wave functions, by the method of Kohn and Sham (1965), which takes the correlation and exchange into account.

The most important virtue of the density-functional method is that electron correlation is accounted for with a computation no more demanding of computer time than the HF method. Further, the computer time required does not increase as rapidly with system size. The method, however, requires appropriate functionals. Great progress was achieved in this respect in the early 1990s (see, for example, Becke 1997). Since that time, DFT has become a standard tool for molecular structure determination. Some of the strategies used are briefly described by Mattsson (2002).

5.2.4 The solvent as dielectric

Recently, both *ab initio* and semiempirical quantum-mechanical treatments have been adapted so as to take into account the role of solvent. Fundamentally, the Schroedinger equation would have to be modified so as to bring the isolated molecule from its state in vacuo to the solution state, that is, to implement the perturbation of the wave function of the molecule by the solvent. Various methods for achieving this goal have been proposed. Representation of the solvent as a continuous dielectric, with a solute contained in a cavity, was pioneered by Born (1920) for ions and by Kirkwood and Onsager for dipolar molecules (Onsager 1936), and implemented as the quantum-Onsager *self-consistent reaction field* (SCRF) method. The electrostatic contribution to the energy for a dipole μ in a spherical cavity of radius a immersed in a medium of relative permittivity ϵ_r is given by eqn 5.4:

$$\Delta U_{\text{elect}} = -\frac{(\epsilon_r - 1)}{8\pi\epsilon_0(\epsilon_r + 1/2)a^3}\mu^2 \tag{5.4}$$

It arises from the *reaction field*, that is, the field within the cavity due to the charges induced in the wall of the cavity by the dipole, acting back on the dipole.

Table 5.1 Calculated (Hartree–Fock, second-order Møller–Plesset) and experimental internal energy of *gauche–anti* rotation of 1,2-dichloroethane *in vacuo* and by quantum–Onsager–SCRF in solution. (After Foresman and Frisch 1993)

Medium	ΔU_{rot} kJ mol^{-1}		
	HF	MP2	Expt
Gas ($\epsilon_r = 1$)	8.2	6.32	5.02
Cyclohexane ($\epsilon_r = 2$)	5.52	4.14	3.81

If the Hamiltonian of the molecule *in vacuo* is $\mathcal{H}_0$, the Hamiltonian of the same molecule in the cavity may be written: $\mathcal{H}_0 + \mathcal{H}_{rf}$, where $\mathcal{H}_{rf}$ is given by eqn 5.5 (Tapia and Goscincki 1975):

$$\mathcal{H}_{rf} = -\mu^{T}\frac{(\epsilon_r - 1)}{(\epsilon_r + 1/2)}\langle\Psi|\mu|\Psi\rangle \tag{5.5}$$

Here μ and μ^{T} are the matrix representation of the dipole moment operator, and its transpose. This calculation can be implemented by available programs, such as Gaussian-92 described by Foresman and Frisch (1993). They give as an example (p. 193) the calculation of ΔU for the *gauche-anti* rotation of 1,2-dichloroethane in the gas phase and in solution in cyclohexane ($\epsilon_r = 2$), using the 6-31 + G(d) basis at both the Hartree–Fock and MP2 levels. Their results are summarized in Table 5.1.

Wong *et al.* (1992) used this approach to calculate the shifts of the tautomeric equilibrium between 2-pyridone (**1**) and 2-hydroxypyridine (**2**) in passing from the gas phase to cyclohexane ($\epsilon_r = 2.0$) and to acetonitrile (35.9). The calculated (and experimental) values for the enol/keto ratio were: gas, 2.9 (3.9); cyclohexane, 0.54 (0.57); acetonitrile, 0.020 (0.007).

Refinements of the quantum-Onsager SCRF method add the higher electric moments: quadrupole, octupole, etc., and use an ellipsoidal rather than a spherical cavity. For molecules that are not even approximately spherical, a cavity following the van der Waals envelope of the molecule (similar to that illustrated in Fig. 5.3) may be used (Miertus *et al.* 1981) at the expense of much more complicated calculation.

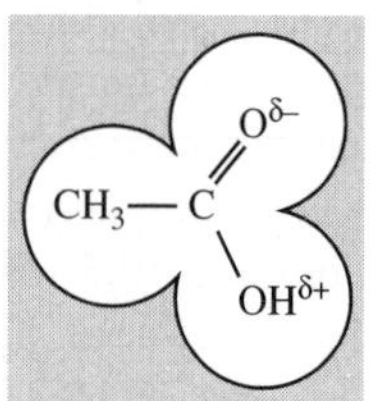

Fig. 5.3 Cavity in a continuous dielectric (schematic) containing an acetic acid molecule.

5.3 Statistical-mechanical methods

In order to calculate quantities that can be compared with experimental measurements, almost all of which relate to systems containing very large numbers of molecules, it is helpful to use statistical methods, and by the formalism of statistical thermodynamics, calculate such properties as molar enthalpies, entropies and Gibbs energies. Two distinct methods are employed: the *Monte Carlo* method, which can yield only equilibrium properties, and

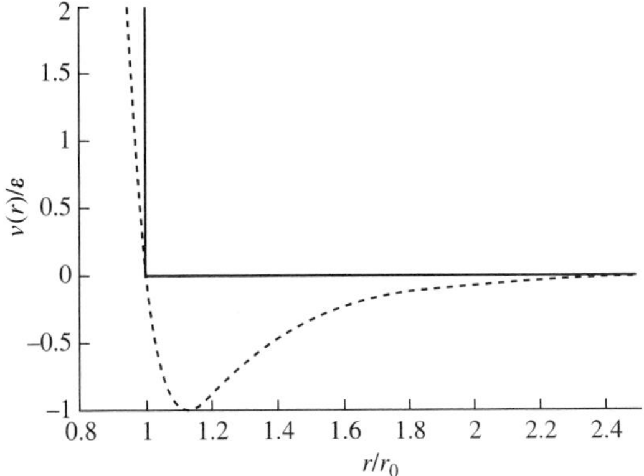

Fig. 5.4 Hard-sphere (solid line) and Lennard–Jones (12–6) potential energy functions.

Molecular Dynamics, which can yield information about both equilibrium properties and time-dependent processes. In these methods, a number in the range 100–10000 of molecules is considered (far too large for quantum mechanics).

The molecules are represented as objects exerting forces on their neighbours (e.g. see Berry *et al.* 2000: chapter 10), so that a potential energy of pairwise interaction may be defined. As was mentioned above, the simplest form of potential energy function usable with atoms is that between hard spheres. A more realistic function, but one which is still reasonably easy to handle mathematically, is the Lennard–Jones potential, a common form of which is represented by eqn 5.6:

$$v(r) = 4\epsilon\left[\left(\frac{\sigma}{r}\right)^{12} - \left(\frac{\sigma}{r}\right)^{6}\right]. \tag{5.6}$$

where ϵ is the depth of the energy minimum and σ is the collision distance, within which the potential begins to be positive. This function and the hard-sphere potential are illustrated in Fig. 5.4. Another is the Buckingham potential, similar to the Lennard–Jones, but with the repulsive r^{-12} term replaced by an exponential in r. It is more realistic, but less easy to deal with in computation.

Modelling polyatomic molecular interactions requires a more complicated function. One approach is through *interaction-site models*, which represent a molecule as an arrangement of sites or centres of force, which are usually the centres of larger atoms or groups such as methyl, each interacting with similar sites on the other molecule, as in eqn 5.7:

$$V(\boldsymbol{r}, \phi, \theta, \psi) = \sum_{i,j} 4\epsilon_{i,j}\left[\left(\frac{\sigma_{i,j}}{r_{i,j}}\right)^{12} - \left(\frac{\sigma_{i,j}}{r_{i,j}}\right)^{6}\right] \tag{5.7}$$

The index i runs over the sites on one molecule, and j over those on the other. The angles ϕ, θ, and ψ are the Euler angles that define the orientation of the second molecule with respect to the first. Other versions of the multi-site model use hard-sphere repulsions at each site, and attractive interactions of

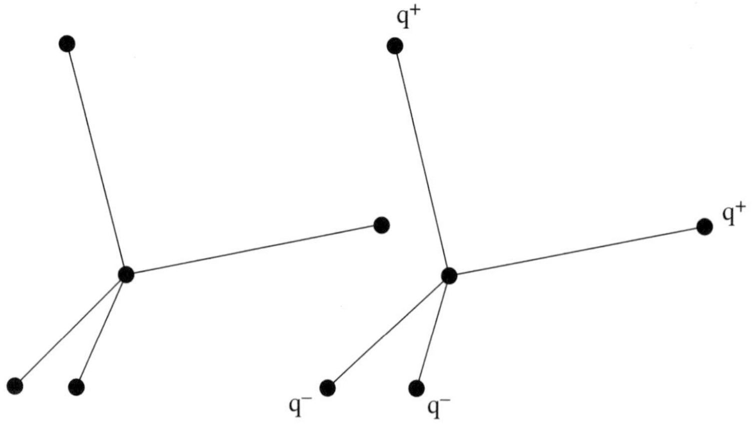

Fig. 5.5 A stereo view of the ST-2 model of water (Stillinger and Rahman 1974). Charges are placed at the tetrahedral angles: $+q$ at 100 pm from the centre representing the proton positions, and $-q$ at 80 pm representing the non-bonding pairs. The electrostatic potential due to these charges is in addition to a Lennard–Jones potential acting between oxygen nuclei as centres.

several kinds. These include the London (dispersion) potential having r^{-6} dependence on intermolecular separation, and if the molecule is polar, longer-range electrostatic interactions arising from charges or dipoles suitably distributed in the molecule. Figure 5.5 depicts one model for the water molecule.

More complex molecules require also that their internal motions: bending, internal rotations, etc., be represented. The molecules are then usually treated as purely classical objects, although modern treatments may also incorporate quantum degrees of freedom in a consistent manner. The Car–Parrinello method, described as *ab-initio molecular dynamics* (Car and Parrinello 1985), is an example.

The number of molecules should be large enough, but too large a number makes the statistical-mechanical calculation too slow. The use of periodic boundary conditions, so that a molecule that passes out of the box on one side reappears on the other, avoids the introduction of spurious boundary effects. Provided the length scales exhibited by the system are smaller than the box size, periodic boundary conditions have little effect on the predicted properties. Thus typical fluids are readily simulated. Restricting the calculation to a 'neighbour list' (only those molecules within a certain distance from a particular molecule) saves time that would be wasted on negligible interactions at long distances (see Fig. 5.6.). The simulation of fluids near their critical points and of ionic liquids, in which long-range interactions are important, require special techniques, such as Ewald summation (Ewald 1921; Leach 1996: 294–8).

5.4 Monte Carlo method

As the name suggests, the *Monte Carlo* method is dependent on *probability*. To calculate the average value of a property G (say, the energy of the system) that can be expressed as a function of all the coordinates of the molecules in a system, $G(\boldsymbol{r})$, it is necessary to evaluate expressions of the type:

$$< G >= \frac{\int G(\boldsymbol{r})p(\boldsymbol{r})\mathrm{d}\boldsymbol{r}}{\int p(\boldsymbol{r})\mathrm{d}\boldsymbol{r}} \tag{5.8}$$

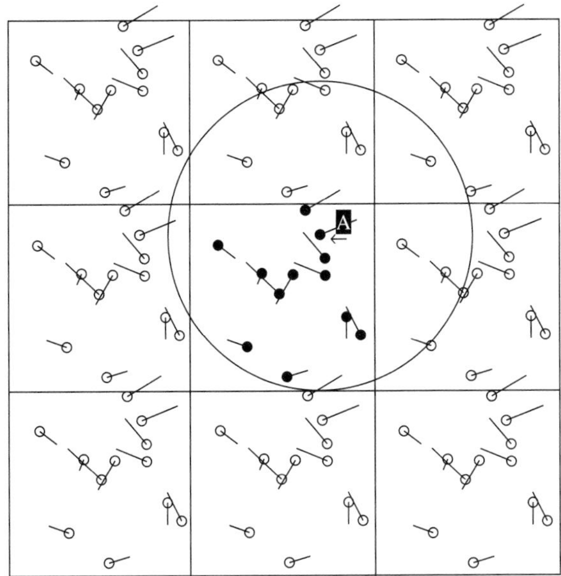

Fig. 5.6 Periodic boundary conditions in two dimensions. 12 molecules; neighbour list boundary shown for molecule 'A'.

where $P(\boldsymbol{r})$ is the weighting (Boltzmann) function, $\exp(-U(\boldsymbol{r})/kT)$. To avoid wasting time on calculations for impossible or highly improbable configurations, the Metropolis *et al.* (1953) method is normally used. Starting from an arbitrary but possible configuration, random changes in coordinates are made, governed by the probability of transition from the present configuration. If the change in energy of the system ΔU will be negative, the new state is accepted. If the change in energy will be positive, the new configuration is accepted only if a random number $0 \leq n \leq 1$ is less than $\exp(-\Delta U/kT)$. Such a series of random steps is called a Markov chain. After many steps, typically from 10^4 to more than 10^6, the mean values of calculated properties will converge to 'equilibrium' values. It may be shown that simple averages over the configurations retained are properly weighted by the Boltzmann factor, and give the correct results, provided that the whole of the configurational space is accessible (the process is 'ergodic'). Failure to reach a portion of configuration space owing to barriers of low probability (high energy) can result in very large errors. Techniques to assure ergodicity include simply lengthening the Markov chain, increasing the number of molecules in the ensemble, starting from a number of different initial configurations, or altering the way in which random steps are made. Unfortunately all these increase the computer time needed. Other methods can be explored, such as the 'hot start', using larger perturbations at first, so the initial configuration is quickly destroyed.

Jorgensen (1982), using 125 molecules and Markov chain lengths 2.5×10^5–7×10^5, obtained for water (at 298 K, 1 atm.) the following (experimental values in parentheses): ΔU_{vap} 41.95 (41.47) kJ mol^{-1}, V_{mol} 18.1 (18.0) cm^3 mol^{-1}, C_{p} 86.5 (75.2) J mol^{-1} K^{-1}, isothermal compressibility 35.5 (46.3) $\times 10^{-6}$ atm^{-1}. Jorgensen's model of the water molecule, designated TIP4P, is a rigid 'ball-and stick' structure, with fractional charges on the **H** and **O** atoms. The repulsive and London attractive potentials are represented by a Lennard–Jones potential centred on the oxygen.

5.5 Molecular dynamics

The method known as *molecular dynamics* begins similarly to the Monte Carlo method, by placing a number of molecules, represented by a suitable charge distribution, in a simulation cell, at time t_0, with assumed initial positions, orientation and velocities. The force acting on each molecule by all the others is then calculated from the gradient of the assumed intermolecular potential. Newton's Second Law of Motion, eqn 5.9 with finite differences replacing the differentials, is then used to calculate the acceleration of each molecule, its new velocity

$$m\frac{d^2r}{dt^2} = \mathbf{F} = -\nabla v(\boldsymbol{r}) \tag{5.9}$$

at time $t_0 + \Delta t$, and its new position resulting from the average velocity over the time interval Δt. The magnitude of the time step size Δt must be chosen carefully, as too small a value will unnecessarily prolong the calculation, while too large a value will result in unphysical configurations. The calculation is repeated for a new step. After each step (or short series of steps), instantaneous values of total energy, pressure, etc., can be calculated. The process continues until the accumulated averages of energy, etc., settle down to equilibrium values.

5.6 Solvation calculations

Going from pure liquids to solutions, by either MC or MD methods, the pure solvent is first simulated, and then one of the solvent molecules in the model is replaced by a solute molecule. After a number of cycles to allow the system to relax to accommodate the intruder, the chain of calculations continues until convergence of the average properties is again achieved. The solvation energy is the difference between the energies of the systems with and without the solute molecule. It is a small difference between two large numbers, so subject to large relative error. Nevertheless results have been obtained. Simkin and Sheikhet (1995) surveyed results of calculations of solvation of a number of hydrocarbons, polar non-electrolytes and electrolytes in water and a few other solvents. The degree of agreement with experimental values is highly variable. Reasons for the sometimes very large errors are discussed.

5.7 Integral equation theories

It can be shown (e.g. see Hansen and McDonald 1986, Section 2.5) that all thermodynamic quantities can be calculated from the radial distribution function, $g(\boldsymbol{r})$. This function, which can be obtained experimentally from

X-ray or neutron diffraction, can also be calculated, provided the intermolecular potential function is known. The aim of the integral equation theories is to calculate the radial distribution function from the intermolecular potential. There are several versions, that differ according to the choice of simplifying assumptions made. They each result in an equation that contains $g(\mathbf{r})$ associated with the Boltzmann factor $\exp(-\beta v(\boldsymbol{r}))$. For example, the PY equation (Percus and Yevick 1958; Hansen and McDonald 1986: 119):

$$\exp[\beta v(\boldsymbol{r})]g(\boldsymbol{r}) = 1 + \rho \int [g(\boldsymbol{r} - \boldsymbol{r}') - 1](1 - \exp[\beta v(\boldsymbol{r}')])g(\boldsymbol{r}')\mathrm{d}\boldsymbol{r}' \quad (5.10)$$

Here β is $1/kT$. The hypernetted chain (HNC) equation (van Leeuven *et al.* 1959) and the equation of Born and Green (1949) involve similar terms, though they are in logarithmic form. The Born–Green equation also contains the first derivative with respect to r, so it is an integro-differential equation for $g(\boldsymbol{r})$ in terms of $v(\boldsymbol{r})$.

5.8 Some results

The whole field of theoretical calculation as it relates to solutions and solvent effects is currently very active. It is not possible in a brief compass to mention all the important advances. We present a random selection of examples of this work, with no particular theme, and without any attempt to be comprehensive.

5.8.1 Microsolvation

By *microsolvation* we mean the association in the gas phase of a single central molecule or ion with a small number of another species, that may be considered a ligand or a potential solvent. For calculation of solvation energies quantum-mechanically, one would like to be able to start with a 'solute' molecule, **A**, and a number of 'solvent' molecules **B**, calculate to a certain level of theory the energy of **A** and **B** alone, then of a series of *supermolecules* or clusters $\mathbf{A} \cdot \mathbf{B}$, $\mathbf{A} \cdot 2\mathbf{B}$, $\mathbf{A} \cdot 3\mathbf{B}$, etc., and a series of clusters $2\mathbf{B}$, $3\mathbf{B}$, etc., until the energy difference represented by:

$$\mathbf{A} + n\mathbf{B} \rightarrow \mathbf{A} \cdot n\mathbf{B} \quad \Delta U_n$$

becomes independent of n. This difference, then, should be the solvation energy of a molecule of **A** in the solvent **B**. In principle one should be able to calculate this energy to any desired degree of accuracy. Unfortunately, to do this by pure quantum-mechanical methods would strain the resources of any presently-available computer, as the number of molecules needed to reach the constant difference would be large. By severely limiting the number of solvent molecules, results can be obtained, however, that may be compared with experimental data on reactions in the gas phase, by mass-spectrometric methods.

Mass-spectrometric studies by Kebarle and co-workers (Arshadi *et al.* 1970; Kebarle 1972) have enabled determination of the enthalpies of successive reactions (in the gas phase at low pressures, in the presence of a gas such as nitrogen or methane, which functions both as a heat bath and as a chemical ionization carrier) of the types:

$$M^+(H_2O)_{n-1}(g) + H_2O(g) = M^+(H_2O)_n(g)$$

and

$$X^-(H_2O)_{n-1}(g) + H_2O(g) = X^-(H_2O)_n(g)$$

where M^+ is H^+ or an alkali-metal ion and X^- is a halide ion. Their results are summarized in Table 5.2. Later work from the same laboratory (Lau *et al.* 1982) refined the values for H^+.

In Fig. 5.7 the stepwise enthalpy changes are plotted against the reciprocal of the number of water molecules after the first in the product ion, $1/(n-1)$. With the exception of the point for $n = 4$, which is above the line, suggesting unusual stability for the ion $H_9O_4^+$ (Berry *et al.* 2000: 309), the values decline regularly, tending towards a value somewhat below the enthalpy of

Table 5.2 Enthalpy changes for gas-phase reactions: $M^+(H_2O)_{n-1} + H_2O = M^+(H_2O)_n$ and $X^-(H_2O)_{n-1} + H_2O = X^-(H_2O)_n$ $-\Delta H_{n-1,\,n}$(kJ mol^{-1})

n	**H$^+$**	**Li$^+$**	**Na$^+$**	**K$^+$**	**Rb$^+$**	**Cs$^+$**	**OH$^-$**	**F$^-$**	**Cl$^-$**	**Br$^-$**	**I$^-$**
1	690	142	100	75	67	57	94	97	55	53	43
2	151	108	83	67	57	52	69	69	53	51	41
3	93	87	66	55	51	47	63	57	49	48	39
4	71	69	58	49	47	44	59	56	46	46	
5	64	58	51	45	44		59	55			
6	54	51	45	42							
7	49										
8	43										

Source: Kebarle (1972), converted to kJ mol^{-1}.

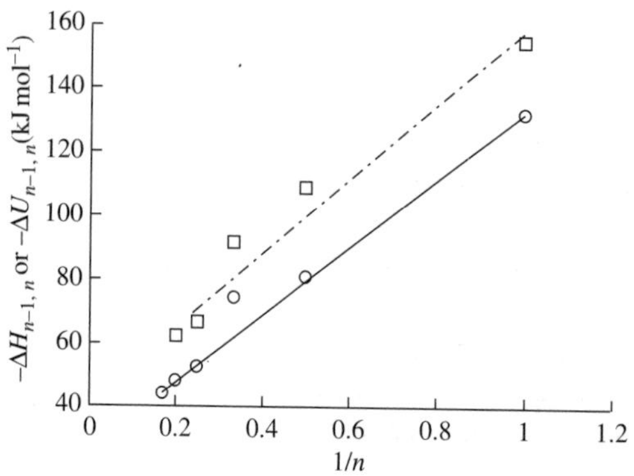

Fig. 5.7 Stepwise experimental, gas-phase enthalpies of hydration of H_3O^+, circles, (Lau *et al.* 1982) and corresponding theoretical energies, squares, (Newton 1982), plotted versus the reciprocal of the number of water molecules. NB, n in this and the following figures corresponds to $n-1$ in Table 5.2.

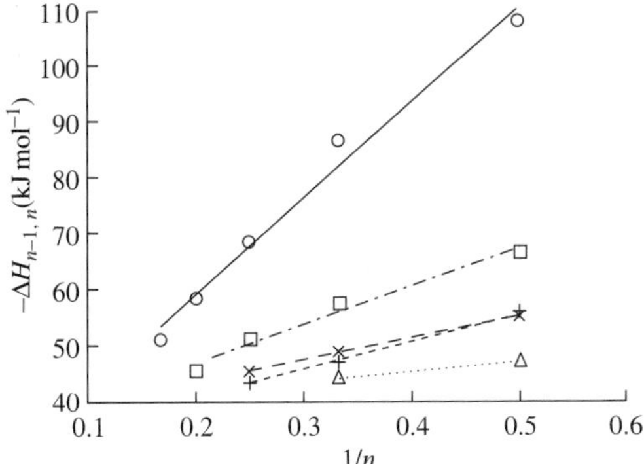

Fig. 5.8 Stepwise experimental, gas-phase enthalpies of hydration of alkali-metal ions (Kebarle 1972) plotted versus the reciprocal of the number of water molecules. Circles, Li^+; squares, Na^+; oblique crosses, K^+; plus signs, Rb^+; triangles Cs^+.

vaporization of bulk water (this is presumably because the molar enthalpy of vaporization from a very small droplet is less than the bulk value). This experimental system is fully analogous to the quantum-mechanical model used by Newton (1977). His *ab-initio* (4–31G) calculated results for ΔU for these reactions are also shown in Fig. 5.7. They also show a marked drop in the magnitude of the energy change for the addition of the fourth water molecule.

The similar hydration of hydroxide ion greatly reduces its energy, an observation that is consonant with the great increase in the chemical activity of OH^- as the concentration of water in a solution of NaOH in mixed DMSO/water solvent is decreased (see Chapter 6).

The successive values of $-\Delta H_{n-1,n}$ for the alkali hydrates recorded in Table 5.2 fall off smoothly, showing no favoured coordination number, CN (Fig. 5.8). Notably, this is true for Li^+ out to $n=6$ (one might have expected CN = 4 to be favoured). The halide ion hydrates show a similar but less steep descent from lower initial values. Fluoride ion is an exception, as the first two water molecules form particularly strong hydrogen bonds with this ion.

Kebarle cites calculations using a model involving r^{-6} attractive and r^{-12} repulsive potential energy terms, plus ion/dipole, ion/induced-dipole and dipole/dipole terms that agreed well with most of the experimental findings for the alkali metal ion hydrates. It was found necessary to allow explicitly for some covalent interaction between sodium or (especially) lithium and the first two or three water molecules.

5.8.2 A classic reaction: The S_N2 reaction, gas-phase versus solution

Probably the most intensively studied reaction, both experimentally and through calculations, is the S_N2 nucleophilic substitution of the general type:

$$Y^- + R\text{-}X \rightarrow Y\text{-}R + X^-$$

These reactions show great sensitivity to the nature of the solvent medium in which they are conducted, in respect of reaction rates, equilibria, stereochemical outcome, and even in the formation of side-products, for example, via elimination. All these different aspects have been subject to theoretical investigations. The reader is referred to the excellent text by Shaik *et al.* (1992) for full discussion of the different theoretical models that have been used, and the derived results.

Of special interest are the energy barriers for given reactions. In principle, calculation of the potential energy surfaces can provide these quantities. In practice, the majority of studies report one-point calculations on SCF-minimized geometries. Generally, calculations for gas-phase S_N2 reactions use the model depicted for the identity reaction:

$$X^- \cdots CH_3{-}X \rightarrow [X{-}CH_3{-}X]^- \rightarrow X{-}CH_3 \cdots X^-$$

C_1 TS C_2

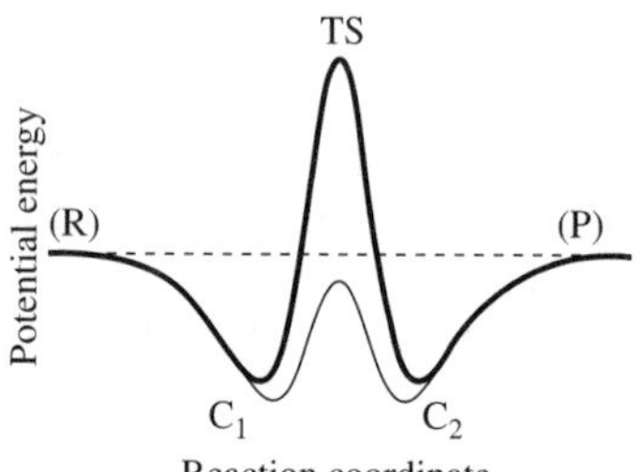

Fig. 5.9 Reaction profile (schematic) for an S_N2 reaction in the gas phase, showing potential wells corresponding to ion–dipole complexes. The energy of the transition state may be higher (heavy curve) or lower (light curve) than the energy of the reactants. Redrawn after Shaik *et al.* (1992), with permission.

The reaction profile for reactants (R) going to products (P), shown schematically in Fig. 5.9, shows a double-well potential, in which the minima C_1 and C_2 represent *ion-dipole complexes*, separated by a central energy barrier representing the transition state (TS) containing a pentacoordinated carbon atom. Computations of the central barrier height above the energy of the well-separated reactants have been performed for a number of gas-phase identity reactions, using *ab-initio* as well as various semiempirical methods: see Table 5.3 (based on Shaik *et al.* 1992: 165, table 5.2). The results for chloride ion demonstrate the difficulty of obtaining precise agreement between methods in such calculations, which involve differences between large numbers.

The major energetic change when one goes from the gas phase to a solution phase, especially in aqueous or alcoholic solvents, is stabilization of the ionic species X^- through H-bonding. The charge-dispersed TS becomes stabilized to a much lesser extent. If the condensed-phase reaction is performed in an aprotic solvent that is, however, highly polar, such as DMF or DMSO, then the ions X^- can no longer be solvated through H-bonding, and will be much less stabilized. This is illustrated in Fig. 5.10, which depicts the results of calculation for the identity reaction:

$$Cl^- + H_3C{-}^*Cl \rightarrow Cl{-}CH_3 + {}^*Cl^-$$

in the gas phase, in water and in DMF.

'Real' S_N2 reactions of interest are, of course, non-identity reactions. Various investigations have reported on central energy barriers and geometries of the transition structures, as well as reaction energies, both for the gas-phase and aqueous reactions. Again, the reader is referred to the text by Shaik *et al.* (1992). The energetics of the reaction of methyl bromide with hydroxide ion microsolvated by 0, 1, 2, 3,... water molecules have been investigated by Bohme and Mackay (1981) using mass spectrometry (See Reichardt's discussion of the energetics of this and related reactions (Reichardt 2003: 155–162).

Table 5.3 Identity S_N2 reactions: Comparison of central barrier heights calculated by various methods[1]

		Ab initio			Semiempirical		
Entry	X	3-21G	4-31G	Large basis sets and CI	MNDO	AM1	INDO(λ)
1	F	51	49	71[2], 75–84[3], ≈88[4], ≈71[5], 71.5[6], 84[7], 54[8]	188	—	91.6
2	Cl	21	23.0	58.2[1], ≈50[4], ≈66[5], 58[9], 53[10], 26[11], 72[8]	44	37.9	91.6
3	Br			53.1[7], 48.5[6]	33.5	23	
4	I					10.0	
5	HOO		77				
6	FO		79				
7	HO		89		201		1.7
8	CH_3O		98		223		
9	HS		65				
10	HCC		210				
11	NC		183		113		
12	CN		119				
13	NH_2		≈160				
14	H		218	255–278[2], 230[12]	−8		216
15	NH_3[13]		78				
16	H_2O[13]			42[14]			

[1] From Shaik *et al.* 1992: 165, table 2. See references therein. Values are converted to kJ mol^{-1}.
[2] 6-31G*//6-31G*
[3] Triple zeta + diffuse functions.
[4] Double zeta + diffuse functions.
[5] Double zeta + diffuse functions + MP4/SDTQCI corrections.
[6] Double zeta = diffuse functions + polarization functions + MRD/CI
[7] Double zeta = diffuse functions + polarization functions.
[8] MP2/6-31 + G* optimization.
[9] 6-31 + G* + MP4//4-31G.
[10] 6-31G* + MP4//4-31G.
[11] 4-31G + MP4//4-31G.
[12] 6-31+ + G** + MP2.
[13] For the neutral nucleophiles, NH_3 and H_2O the reaction is: X: + CH_3–X^+ = X–$CH_3{}^+$ + X:
[14] 6-31G** + MP3//3-21G including ZPE correction.

5.8.3 Solvatochromism: Theoretical calculations

Several authors have carried out theoretical calculations of the solvation of molecules that form the bases of solvatochromic scales of solvent properties. To take one example, we may consider the molecule 4-[(4′-hydroxyphenyl)azo]-*N*-methylpyridine (**3**) (called 'Buncel's dye' by Rauhut *et al.* 1993), which exists as a resonance hybrid of two structures. A UV-VIS spectroscopic study of **3** and related substances revealed a strong solvatochromic effect, which served as the basis of the establishment of a π^*_{azo} solvent polarity scale (Buncel and Rajagopal 1989, 1990, 1991). The theoretical study of Rauhut *et al.* (1993), was based on AM1 methodology (Dewar and Storch 1985) but used a double electrostatic reaction field in a cavity, dependent on both the

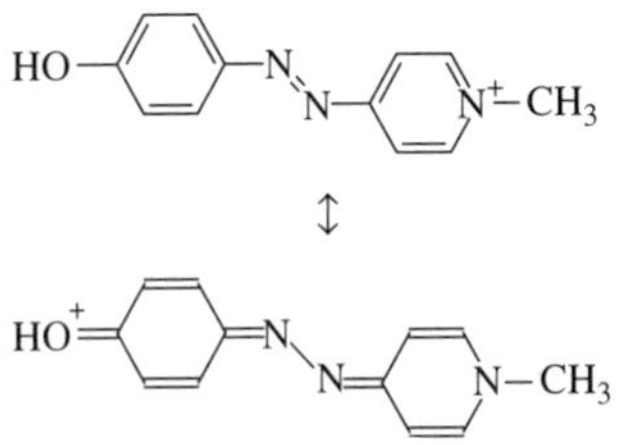

HO–C₆H₄–N=N–C₅H₄N⁺–CH₃
↕
HO⁺=C₆H₄=N–N=C₅H₄N–CH₃

3

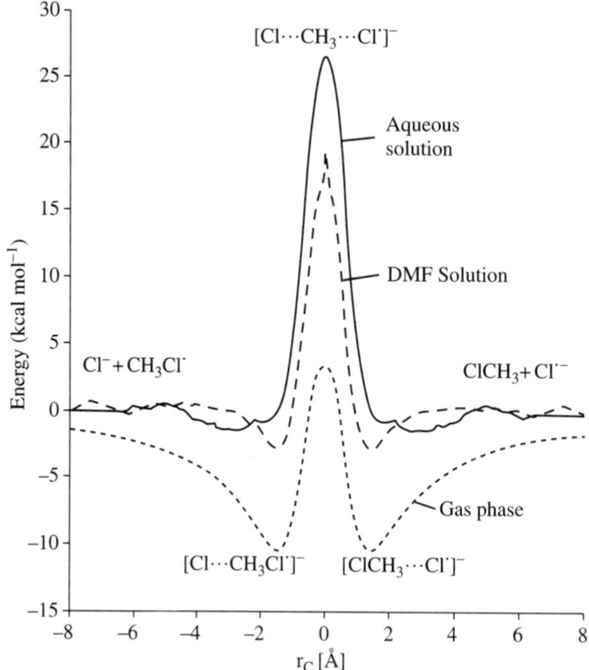

Fig. 5.10 Calculated internal energies in the gas phase (short dashes) and the potential of mean force in DMF (long dashes) and in aqueous solution (solid curve) for the reaction of Cl^- with CH_3Cl as a function of the reaction coordinate, r_c, in angstroms. Reproduced from Chandrasekhar and Jorgensen (1985), with permission.

dielectric constant and the refractive index. Nuclear motions interact with the medium through the dielectric constant, but electronic motions are too fast; only the extreme high-frequency part of the dielectric constant is relevant. These authors were able to evaluate solvent-specific dispersion contributions to the solvation energy. The calculations reproduced satisfactorily the experimental solvatochromic results for **3** in 29 different solvents. The method has also been successfully applied to other solvatochromic dyes, including Reichardt's $E_T(30)$ betaine.

Matyushov *et al.* (1997) calculated the energy changes that form the bases of the π^* and $E_T(30)$ scales, using a model in which the solvent molecules were represented as hard spheres and the solute (dye) molecules as hard hemisphere-capped cylinders, and only relatively long-range solute-solvent interactions were considered. Interactions considered were permanent dipole–dipole (*perm*), London dispersion (*disp*), and dipole–induced-dipole (*ind*). The polarizabilities were assumed to be isotropic. Both π^* and $E_T(30)$ were shown to contain contributions from all three interactions. The interesting result emerged that while both scales are similarly affected by *disp* and *perm*, the *ind* term reinforces the other two in the case of π^*, but acts in the contrary sense with $E_T(30)$. The reason adduced is that in passing from the ground to the excited state, the dipole moment and the polarizability of 4-nitroanisole (taken as the type dye for π^*) both increase, whereas with the Reichardt betaine-(30) the polarizability increases (greatly) but the dipole moment decreases. The agreement between the experimental and calculated values of both parameters is good, except that π^* is overestimated for the more polarizable chlorinated and aromatic solvents, while $E_T(30)$ is underestimated for the alcohols. The latter discrepancy is undoubtedly due to the explicit neglect of hydrogen bonding, and is in keeping with the observation that $E_T(30)$ falls between the dipolar and acidic groups of parameters in Fig. 4.2.

Among the applications of integral equation theory to solutions, Koga and Tanaka (1996) used a reference interaction-site model (RISM), an approximation due to Chandler and Andersen (1972), to calculate the local solvation behaviour of naphthalene in supercritical carbon dioxide. They used a ten-site model for naphthalene, and a three-site model for carbon dioxide, in which partial charges were placed on the sites in addition to Lennard–Jones potentials acting between sites on different molecules. Their potential function was a sum over the 10 naphthalene sites and the three carbon dioxide sites of terms of the form of eqn 5.11:

$$v_{i,j} = 4\epsilon_{i,j}\left[\left(\frac{\sigma_{i,j}}{r_{i,j}}\right)^{12} - \left(\frac{\sigma_{i,j}}{r_{i,j}}\right)^{6}\right] + \left(\frac{z_i z_j e^2}{r_{i,j}}\right) \tag{5.11}$$

They demonstrated that the RISM integral equation and MC methods gave comparable results, and that the nearest CO_2 molecules showed preferred orientations in certain positions: parallel to the naphthalene molecular plane and to the long axis when above the plane, but perpendicular to the plane whether 'abeam' or 'ahead' of the naphthalene. They also obtained the partial molal volume of naphthalene as a function of the density of solvent; it is negative at low densities, increasing to reasonable positive values as density increases.

In a study of aqueous solvation of *cis* and *trans-N*-methylacetamide, Yu *et al.* (1991) compared MC, MD, and several integral equation methods with each other and with experimental data. It was apparent that the various integral equation methods differed in their applicability to this system. The HNC method in particular, though recommended by some for use with polar substances, gave poor results for the absolute value of the solvation free energy for both conformers (the wrong sign, owing to overestimation of the entropy contribution), though the difference between them was close to the experimental value.

5.9 Hydrophobic solvation

The hydrophobic effect, mentioned in Section 2.7, has received much attention. Némethy and Scheraga (1962) were able to demonstrate that the decrease in entropy accompanying the entry of a hydrocarbon into aqueous solution was due to increased hydrogen bonding among the water molecules in the layer nearest the solute molecule. Okazaki *et al.* (1979), whose work was mentioned in Chapter 2, obtained for the transfer functions of methane from the gas phase to water at 25 °C: $\delta A = 0.8$ (almost identical with the experimental value base on the solubility 2.48×10^{-5} mole fraction), $\delta U = -1.7$ and $T\delta S = -2.5$ (all in kJ mol^{-1}), showing that, while there is a decrease in energy from the increased H-bonding, the loss of entropy dominates the solubility. They later found (Okazaki *et al.* 1981) the trend in stabilization energy as the size of the solute increases to be not monotonic: ethane was found to be less stabilized than methane, but pentane more stabilized, and they noted some changes in structural details. Abraham (1982) found, in the course of a statistical analysis of the experimental solubilities and their temperature dependence of a large number of non-polar gases in many solvents, that for normal hydrocarbons the hydration of the terminal methyl groups is entropy dominated, but that for the methylene groups is it enthalpy dominated. Tanaka (1987), as a result of a study by MC and by the RISM integral equation method, reported that for non-polar solutes of increasing size, H-bonding increased, with resulting increase of solubility.

Other workers have used various methods (MC, MD, integral equations) to investigate thermodynamic aspects of hydrophobic solvation (Abraham 1982; Guillot *et al.* 1991; Durrell and Wallqvist 1996; Cann and Patey 1997). Some have examined effects of solute size or curvature (Okazaki *et al.* 1981; Tanaka 1987; Ashbaugh and Paulaitis 1996; Chau *et al.* 1996). Others have focussed on specific aspects such as heat capacity (Madan and Sharp 1996), volume and compressibility (Matubayasi and Levy 1996).

For a final example, Guillot *et al.* (1991), in a series of MD simulations of solutions of the rare gases and methane in water and in methanol, showed that the solvent molecules in the first shell are tangentially oriented (as depicted in Fig. 2.4(b)). Figure 5.11 shows the pair distribution functions for argon in water: $g_{AO}(r)$ for the distribution of oxygen and $g_{AH}(r)$ for the distribution of hydrogen, near argon. The first maxima in the two distributions are nearly coincident. Meng and Kollman (1996) put forward Fig. 5.12, which shows the same phenomenon in the distribution of O and H around methane in water. Guillot and Guissani (1993) also showed that the order of solubilities: Ne < Ar, etc., is dominated by ΔU, and that ΔS values are temperature-dependent in a manner that gives rise to minima in some of the solubilities as functions of temperature. Bagno (1998) presented a diagram (reproduced as

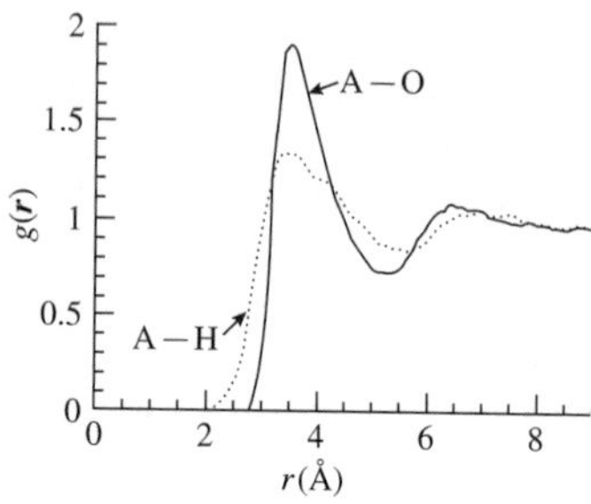

Fig. 5.11 Calculated pair distribution functions for argon in water: argon-oxygen $g_{Ar\text{-}O}$ and argon-hydrogen $g_{Ar\text{-}H}$. Note near coincidence of the first maxima (Guillot *et al.* 1991). Reprinted with permission from *The Journal of Chemical Physics* **95**, Copyright 1991, American Institute of Physics, and the authors.

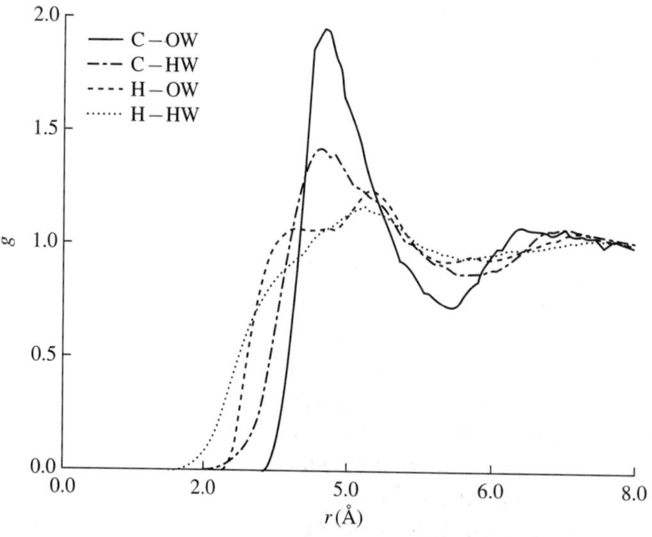

Fig. 5.12 Calculated pair distribution functions for methane in water: carbon–oxygen $g_{C\text{-}O}$, carbon–hydrogen(water) $g_{C\text{-}Hw}$, hydrogen(methane)–hydrogen(water) $g_{Hm\text{-}Hw}$, and hydrogen (methane)–oxygen $g_{Hm\text{-}O}$. Note the near coincidence of the first maxima for C–Hw and C–O. Reproduced with permission from Meng and Kollman (1996).

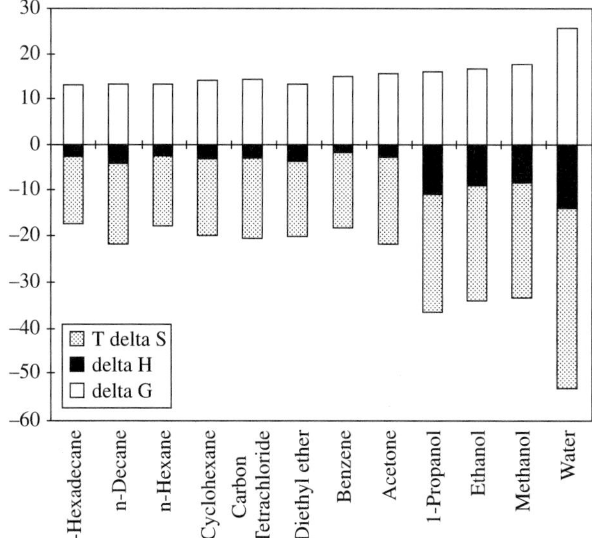

Fig. 5.13 Enthalpy and entropy contributions to the Gibbs energy of solvation of methane in a number of solvents. Reproduced with permission from Bagno (1998).

Fig. 5.13) showing clearly that in all solvents examined, the enthalpy and entropy of solvation of methane are both negative, and that at ordinary temperatures the $T\Delta S$ term is substantially larger than the ΔH term (so the solubility may be described as 'entropy-dominated'). In the hydroxylic solvents, and especially in water, both terms are large, the entropy remaining dominant.

It appears from these studies that the picture of an 'iceberg' around the solute in hydroxylic solvents is overdrawn. The shell of tangentially oriented molecules is only a monolayer, and though these molecules are restricted in their radial motion (giving rise to the entropy loss), they are not strongly ordered within the monolayer.

Problem

1. Prove that the product of any two Gaussians in three dimensions is also a Gaussian. Since a coordinate system is arbitrary, it will suffice to set up the original two on the x-axis, at suitable distances on either side of the origin. The result is particularly simple if they are centred at $-x_1$ and $+x_2$, with $x_1/x_2 = s_1^2/s_2^2$, where s represents the standard deviation.

6 Dipolar aprotic solvents

6.1 Introduction

Dipolar aprotic solvents came into prominence in the 1960s, with the discovery that the rates of a number of reactions of the S_N2 type, and some other reactions as well, were very greatly enhanced, by factors up to 10^9, when carried out in solvents such as dimethylformamide and dimethyl sulphoxide as compared to common hydroxylic solvents such as water or ethanol. This was the pioneering work of Parker (1962, 1967, 1969), Ritchie (1969), Cram (1965) and others (Coetzee and Ritchie 1969). A list of the more common dipolar aprotic solvents together with some of their important physical characteristics is given in Table 6.1.

It is immediately obvious on glancing at the entries in Table 6.1 (all organic molecules) that all these solvents actually contain hydrogen atoms, albeit these are bound to carbon atoms, so the term 'aprotic' is actually a misnomer. In principle, virtually all C–H containing compounds can be deprotonated when in the presence of a sufficiently strong base, so the solvents listed in Table 6.1 can be termed carbon acids, though mostly very weak carbon acids. Importantly, unlike the typical 'normal' acids in which the ionizable hydrogen is bound to the electronegative atom, oxygen, and which are therefore strong hydrogen-bond donors (i.e. HBD solvents), the C–H acids are extremely poor hydrogen-bond donors. Therefore, an alternative terminology for the dipolar aprotic solvents which has been proposed is *dipolar non-HBD* solvents (Mashima *et al.* 1984). However, in the present account the more common naming will generally be retained.

As shown in Table 6. 1, the distinguishing features of the dipolar aprotic solvents are the relatively high dielectric constants ($\epsilon > 15$), dipole moments ($\mu > 5$) and E_T^N solvent parameters ($E_T^N > 0.3$).

Clearly, anions will be poorly solvated in the dipolar aprotic solvents. On the other hand, due to the presence of atoms with lone pairs of electrons, their interaction with cations will be quite effective. It is this characteristic inability to solvate anions, contrasting with cations, which is used to advantage in reactivity studies to be described below.

A dipolar aprotic solvent *par excellence* is dimethyl sulphoxide (DMSO) and most representative studies of solvent effects on reactivities have included DMSO (Buncel and Wilson 1977a). Accordingly, in the following account, DMSO will be highlighted as the prototypical dipolar aprotic solvent, though comparisons with other dipolar aprotic solvents will also be presented.

6.2 Acidities in DMSO and the H_- scale in DMSO–H_2O mixtures

DMSO has been found to be highly suitable for study of ionizations of extremely weak acids. Its conjugate base, $CH_3SOCH_2^-$, or *dimsyl anion*

Table 6.1 Physical properties of some dipolar aprotic non-HBD solvents in order of increasing dipole moment

Name	Formula	bp(°C)	Dielectric constant, ϵ_r	Dipole moment, (10^{-30} C m)	E_T^N	pK_{auto}
Acetone	CH_3COCH_3	56	20.6	9	0.355	32.5
Formamide	$HCONH_2$	210	111	11.2	0.799	16.8
Nitromethane	CH_3NO_2	101	35.9	11.9	0.481	
N,N-dimethylacetamide	CH_3CONMe_2	166	37.8	12.4	0.401	24
N,N-dimethylformamide (DMF)	$HCONMe_2$	153	36.7	13	0.404	29.4
N-methylpyrrolidin-2-one (NMP)	$CH_3NCOCH_2CH_2CH_2$ (ring)	202	32.2	13.6	0.355	24.2
Acetonitrile	CH_3CN	81.6	35.9	11.8	0.46	≥33.3
Dimethyl sulphoxide (DMSO)	CH_3SOCH_3	189	46.5	13.5	0.444	33.3
N,N-dimethylpropylene urea (DMPU)	$CH_2CH_2CH_2N(CH_3)CONCH_3$ (ring)	230	36.1	14.1	0.352	—
Sulpholane	$(CH_2)_5SO_2$	287	43.3	16	0.41	25.5
Propylene carbonate (PC)	$OCH(CH_3)CH_2OCO$ (ring)	242	64.9	16.5	0.491	—
Hexamethylphosphoric triamide (HMPT)	$(Me_2N)_3PO$	233	29.6	18.5	0.315	20.6

(Buncel *et al.* 2003), is stable in DMSO and the glass electrode can be used for potentiometric measurement in DMSO (Terrier *et al.* 1991, 1995). Thus, an acidity scale has been established in DMSO and Table 6.2 records some pK(DMSO) data together with the corresponding pK(H_2O) results (Bordwell 1988). It is noteworthy that for H_2O, pK(DMSO) is of the order of 32, that is it is an extremely weak acid! This, of course, corresponds to OH^- being an extremely strong base in DMSO. Interestingly, relative acidities in DMSO resemble those in the gas phase, more so than in aqueous solution, pointing to the all-important influence of hydrogen bonding in aqueous medium (Pellerite and Brauman 1980; Taft and Bordwell 1988).

DMSO–H_2O mixtures also impart a strong basicity to hydroxide ion. Quantitatively, this is expressed by means of the H_- acidity function, that is:

$$H_- = -\log\left(\frac{a_{H^+} f_A}{f_{HA}}\right). \tag{6.1}$$

As in the case of H_o (see Chapter 3), in dilute aqueous solution $H_- = pH$, since in that state $a_{H^+} = C_{H^+}$ and $f_A = f_{HA} = 1$. A selection of H_- data in DMSO–H_2O, and in several other aqueous dipolar aprotic solvent mixtures, is given in Table 6.3 (Buncel 1975b). The extremely sharp rise of H_- as the last small amounts of water are removed is quite striking. A physical interpretation of this can be given in terms of hydrogen-bonding between

Table 6.2 Equilibrium acidities in dimethyl sulphoxide and in water[1]

Acid	$pK_a(H_2O)$	$pK_a(DMSO)$
F_3CSO_3H	−14	0.3
HBr	−9	0.9
HCl	−8	1.8
CH_3SO_3H	−0.6	1.6
HF	3.2	15
$PhCO_2H$	4.25	11.1
CH_3CO_2H	4.75	12.3
$PhNH_3^+$	4.6	3.6
PhSH	6.61	10.3
NH_4^+	9.2	10.5
CH_3NO_2	10	17.2
PhOH	10.2	18
$CH_2(CN)_2$	11	11
F_3CCH_2OH	12.4	23.6
CH_3OH	15.5	29
H_2O	15.7	32

[1] Bordwell 1988; Stewart 1985.

Table 6.3 Selected H_ data for aqueous binary mixtures with several dipolar aprotic solvents, each with 0.011 M tetramethylammonium hydroxide

Mole% dipolar aprotic component	Aqueous pyridine	Aqueous tetramethylene sulphone	Aqueous dimethyl-formamide	Aqueous dimethyl sulphoxide
20	13.75	13.22	14.20	14.48
40	14.76	14.25	15.75	16.50
60	15.31	15.56	17.34	18.50
80.78				20.68
90.07				21.98
99.59				26.59

Source: Buncel, 1975b.

DMSO and H_2O molecules, that is

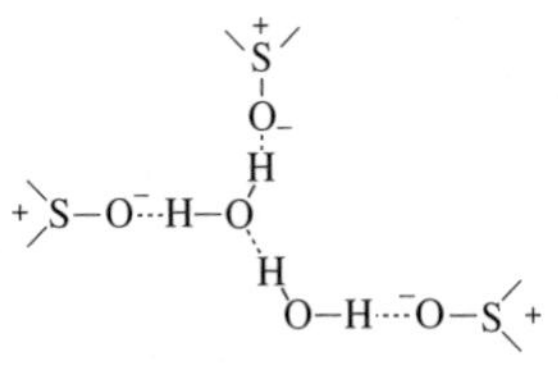

$$\begin{array}{l} \qquad\qquad \overset{+}{S} \\ \qquad\qquad | \\ \qquad\qquad O^- \\ \qquad\qquad \vdots \\ \qquad\qquad H \\ {}^{+}S{-}O^{-}\cdots H{-}O \\ \qquad\qquad \vdots \\ \qquad\qquad H \\ \qquad\qquad O{-}H\cdots {}^{-}O{-}S^{+} \end{array}$$

This H-bonding effectively removes any free H_2O molecules from the bulk solvent, and hence from solvating OH^-, which then becomes correspondingly more effective in removing a proton from a solute in an equilibrium or rate process.

6.3 Use of thermodynamic transfer functions

Approaching solvent effects from the viewpoint of thermodynamic transfer functions allows one to examine in a systematic manner the outcome of medium change, from a protic to a dipolar aprotic reaction medium, in terms of structure and charge distribution in reactants, transition states and products (Buncel and Wilson 1979, 1980).

The free energy of activation for a reaction in a particular standard solvent O, and in a solvent S, may be expressed by eqns 6.2 and 6.3

$$\Delta G_{\mathrm{O}}^{\ddagger} = G_{\mathrm{O}}^{\mathrm{T}} - G_{\mathrm{O}}^{\mathrm{R}} \tag{6.2}$$

$$\Delta G_{\mathrm{S}}^{\ddagger} = G_{\mathrm{S}}^{\mathrm{T}} - G_{\mathrm{S}}^{\mathrm{R}} \tag{6.3}$$

$$\Delta G_{\mathrm{S}}^{\ddagger} - \Delta G_{\mathrm{O}}^{\ddagger} = \left(G_{\mathrm{S}}^{\mathrm{T}} - G_{\mathrm{O}}^{\mathrm{T}}\right) - \left(G_{\mathrm{S}}^{\mathrm{R}} - G_{\mathrm{O}}^{\mathrm{R}}\right) \tag{6.4}$$

where T refers to the transition state and R to the reactants. Simple subtraction yields eqn 6.4. If the standard free energy of transfer between solvents O and S is defined according to eqn 6.5, then eqn 6.6 can be derived from eqn 6.4. A pictorial representation of these relationships is shown in Fig. 6.1.

$$\delta G_{\mathrm{tr}}^{\mathrm{i}} = G_{\mathrm{S}}^{\mathrm{i}} - G_{\mathrm{O}}^{\mathrm{i}} \tag{6.5}$$

$$\delta G_{\mathrm{tr}}^{\mathrm{T}} = \delta G_{\mathrm{tr}}^{\mathrm{R}} + \Delta G_{\mathrm{S}}^{\ddagger} - \Delta G_{\mathrm{O}}^{\ddagger} = \delta G_{\mathrm{tr}}^{\mathrm{R}} + \delta \Delta G^{\ddagger}. \tag{6.6}$$

From eqn 6.6 it is apparent that $\delta G_{\mathrm{tr}}^{\mathrm{T}}$ can be evaluated from calculated values of the transfer free energies of stable solute species, $\delta G_{\mathrm{tr}}^{\mathrm{R}}$, in conjunction with the measured kinetic activation parameters, $\delta \Delta G^{\ddagger}$. The required transfer free energies $\delta G_{\mathrm{tr}}^{\mathrm{R}}$ can readily be obtained from activity coefficient measurements using eqn 6.7

$$\delta G_{\mathrm{tr}}^{\mathrm{i}} = -RT \ln \frac{\gamma_{\mathrm{S}}}{\gamma_{\mathrm{O}}} \tag{6.7}$$

where γ refers to solute activity coefficients in the different solvents (γ_{S} and γ_{O} are referred to the same standard state in solvents S, O, or any other medium). Methods used to obtain these activity coefficients have included vapour pressure, solubility, and distribution coefficient measurements.

By analogy it is apparent that for the equilibrium situation the transfer free energy relationship will be given by eqn 6.8

$$\delta \Delta G_{\mathrm{tr}} = \delta G_{\mathrm{tr}}^{\mathrm{P}} - \Delta G_{\mathrm{tr}}^{\mathrm{R}} \tag{6.8}$$

where $\delta G_{\mathrm{tr}}^{\mathrm{P}}$ is the transfer energy of the products and $\delta \Delta G_{\mathrm{tr}}$ is the difference in the standard free energies of reaction between the two solvents. Transfer functions can also be defined for the other thermodynamic state functions. Since enthalpy changes are often conveniently measurable, the transfer

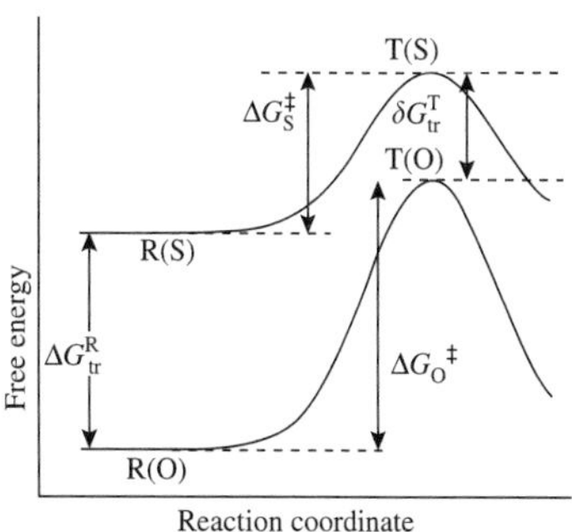

Fig. 6.1 Illustration of relationship between the transfer free energies of reactants and the transition state, and the free energies of activation , for a reaction occurring in two solvent systems. Redrawn from Buncel and Wilson 1980, with permission.

enthalpy, $\delta H^{\mathrm{i}}_{\mathrm{tr}}$, is perhaps the most widely used function. From the second law of thermodynamics, the transfer entropy function is given by eqn 6.9.

$$\delta G^{\mathrm{i}}_{\mathrm{tr}} = \delta H^{\mathrm{i}}_{\mathrm{tr}} - T\delta S^{\mathrm{i}}_{\mathrm{tr}} \tag{6.9}$$

Thus, if both transfer free energies and enthalpies are available it should be possible to achieve complete dissection of the effect of solvent on the various thermodynamic parameters.

Other potentially useful transfer functions are the transfer heat capacity, $\delta C^{\mathrm{i}}_{\mathrm{tr}}$, the transfer volume, $\delta V^{\mathrm{i}}_{\mathrm{tr}}$, and the transfer internal energy, $\delta U^{\mathrm{i}}_{\mathrm{tr}}$ (Abraham 1974). At present, measurements of these transfer functions are available for only a few systems, but they can in certain instances be predicted using scaled-particle theory (Desrosiers and Desnoyers 1976).

On the other hand, use of activity coefficients of transfer, ${}^{\mathrm{O}}\gamma^{\mathrm{S}}$, instead of free energies of transfer, is an equivalent approach for examination of medium effects on reaction rates and equilibria. Parker has emphasized this alternative approach and the reader is referred to his writings, especially for medium effects in bimolecular nucleophilic substitution (Parker 1969).

In Table 6.4 are presented some of the currently available $\delta\Delta G_{\mathrm{tr}}$ data for transfer of ionic species from water to various dipolar aprotic solvents (Cox 1973). The most noticeable feature of these data is the striking increase in the free energy of small anions with high charge density on transfer to the dipolar aprotic solvents. This is consistent with the large rate increase observed in the reactions involving these anions. If the anion is small, it should be a strong

Table 6.4 Free energies of transfer of ions ($\delta\Delta G_{\mathrm{tr}}$) from water to non-aqueous solvents at 25 °C (molar scale, in kcal mol^{-1})

Ion	MeOH	HCONH$_2$	NMeF	DMF	DMSO	Me$_2$CO	PC	MeCN
H^+	2.6	—	—	−3.4	−4.5	−0.7	—	11.1
Li^+	1.0	−2.3	−3.5	−5.3	−3.5	—	5.3	7.1
Na^+	2.0	−1.9	−1.9	−2.5	−3.3	—	2.6	3.3
K^+	2/4	−1.5	−2.0	−2.3	−2.9	0.7	0.8	1.9
Rb^+	2.4	−1.3	−1.8	−2.4	−2.6	0.5	−1.3	1.6
Cs^+	2.3	−1.8	−1.7	−2.2	−3.0	0.4	−3.5	1.2
Ag^+	1.8	−3.7	—	−4.1	−8.0	1.5	3.3	−5.2
Tl^+	1.0	—	—	−2.8	−6.0	—	2.0	2.2
Et_4N^+	0.2	—	—	−2.0	−1.2	—	—	−2.1
Ph_4As^+	−5.6	−5.7	—	9.1	−8.8	−7.1	−8.5	−7.8
Cl^-	3.0	3.3	4.9	11.0	9.2	14.0	10.1	10.1
Br^-	2.7	2.7	3.6	7.2	6.1	10.5	7.8	7.6
I^-	1.6	1.8	—	4.5	3.2	6.7	4.6	4.5
N_3^-	2.5	2.9	—	8.2	5.7	10.9	7.1	7.0
ClO_4^-	1.4	—	0.4	—	—	3.6	—	1.1
OAc^-	3.7	—	—	14.8	11.1	—	—	13.4
BPh_4^-	−5.6	−5.7	—	−9.1	-8.8	−7.1	−8.5	−7.8

Data from Cox 1973, calculated using the assumption, $\delta\Delta G_{\mathrm{tr}}(Ph_4As^+) = \delta\Delta G_{\mathrm{tr}}(BPh_4^-)$.
Abbreviations: NMeF: *N*-methylformamide; PC: propylene carbonate.

hydrogen bond acceptor in protic solvents. Such interactions are absent in DMSO, which accounts for these anions being much less solvated in DMSO than in water. That could also explain the observation (Wooley and Hepler 1972) that the ionization product of water decreases rapidly with increasing DMSO content in mixtures of DMSO and water. In contrast, the free energies for large polarizable anions usually decrease on transfer to dipolar aprotic solvents, as strikingly illustrated for the case of BPh_4^-.

The results for the free energies of transfer of cations do not present as uniform a picture in the sense that the solvents acetone, propylene carbonate, and acetonitrile do not display the same behaviour as solvents such as DMSO and DMF. It is clear, however, that cations are better solvated in the latter solvents than in water.

6.4 Classification of rate profile–medium effect reaction types

By combination of thermodynamic and kinetic measurements one can obtain values for δG_{tr}^{T}, which in conjunction with δG_{tr}^{R} values can be used to pinpoint the cause of rate change with changing solvent. A number of different types can be envisaged. The δG_{tr}^{T}, δG_{tr}^{R} terms can be positive (destabilization), negative (stabilization), or zero (no effect). When both terms have the same sign we call it a *balancing* situation, and with opposite sign a *reinforcing* situation. The rate effects are expected to be largest in the reinforcing situation and smallest in the balancing situation.

In Table 6.5 we summarize the various possibilities and identify the reaction types in terms of positive (negative), initial-state (transition-state) control. For example, case 3 can be described as a *positive transition-state control* reaction type, and so on. Not all of these situations have been observed so far. It would be of interest to design appropriate reactions to complete this classification.

Table 6.5 Transfer free energies of reactants (δG_{tr}^{R}) and transition states (δG_{tr}^{T}) and solvent effects on reaction rates. Classification of reaction types

Case	δG_{tr}^{R}	δG_{tr}^{T}	Effect on rate[a]	Reaction type
1	−ve	−ve	+, 0, or −	Balanced
2	+ve	−ve	+	Positively reinforced
3	0	−ve	+	Positive transition-state control
4	−ve	0	−	Negative initial-state control
5	+ve	0	+	Positive initial−state control
6	0	0	0	Solvent independent
7	−ve	+ve	−	Negatively reinforced
8	+ve	+ve	+, 0, or −	Balanced
9	0	+ve	−	Negative transition-state control

[a] The plus sign refers to rate acceleration, the minus to rate retardation, and zero to no effect.

Source: Buncel and Wilson 1980.

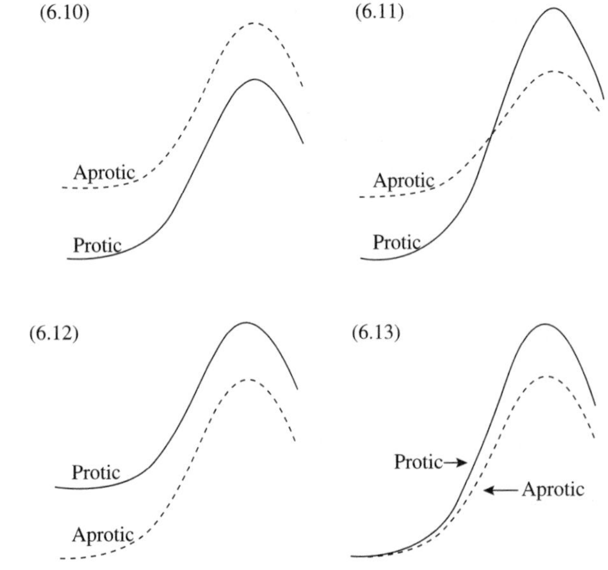

Fig. 6.2 Reaction profile illustrations of the solvent effects on the rates of some representative reactions for changes from protic to dipolar aprotic media, showing enthalpy as the ordinate. Using the classification in Table 6.5, cases (6.10) and (6.12) represent balancing situations, (6.11) positively reinforced, (6.13) negative transition-state control. Redrawn from Buncel and Wilson (1979), with permission.

(6.10) $HO^- + D_2$ (H_2O versus aq.DMSO)

(6.11) O^--C_6H_4-NO_2 + CH_3I (MeOH versus DMF)

(6.12) SCN^- + I-$C_6H_3(NO_2)$-Me (MeOH versus DMF)

(6.13) pyridine + CH_2Cl-C_6H_4-CH_3 (MeOH versus DMF)

In Fig. 6.2 are shown a few examples of reactions (Buncel and Symons 1976; Haberfield 1971) under these types, some of which will be referred to again subsequently. It may be noted that only in example 6.13 is the ground-state effect zero. In the general case, however, the rationalization of medium effects on reaction rates requires knowledge of the pertinent transfer function data for the reactants; these are now becoming increasingly available for neutral molecules as well as ionic reagents. In the case of transfer functions for ionic reagents it has been customary to estimate thermodynamic quantities for single ions by use of *extrathermodynamic assumptions*, such as $\delta\Delta G_{tr}(Ph_4P^+) = \delta\Delta G_{tr}(BPh_4^-)$. These methods are open to criticism, but in certain instances they can be avoided by use of appropriate thermochemical cycles (Gold 1976).

Analogous considerations are applicable in principle to certain excited-state processes. The blue shift of $n \rightarrow \pi^*$ transitions on going to more polar solvents has been analysed in this manner by Haberfield *et al.* (1977) for a number of ketones and azo compounds. According to the terminology given in Table 6.5, with substitution of the excited state for the transition state, the commonly observed blue shift for ketones on transfer from a non-polar solvent to a hydrogen bonding solvent falls under the category of a *negatively reinforced* type of process.

6.5 Bimolecular nucleophilic substitution

The relative inability of dipolar aprotic solvents to interact effectively with (solvate) anions, especially small (hard) anions in which the negative charge is located on an electronegative atom, results in very large rate accelerations in these non-HBD solvents relative to protic ones. According to the terminology in Table 6.5, these processes can be classified as *positively reinforced* since along with destabilization of the reactants there is stabilization of the transition state due to charge dispersal (Hughes–Ingold rules, Ingold 1969).

The above ideas are nicely illustrated by the results in Table 6.6, corresponding to reactions 6.14–6.17 (Reichardt 2003: 249). The first reaction is a simple S_N2 type halide interchange process, the second is similar but involves azide ion as a very effective nucleophile. The last two are aromatic nucleophilic substitution (S_NAr) reactions and are shown as proceeding by way of *sigma complex intermediates* (Buncel *et al.* 1995). Note that eqn 6.17, which involves a neutral nucleophile is much less sensitive to solvent change than the former three which all involve anionic nucleophiles; this is in accord with expectations.

$$Cl^- + CH_3{-}I \overset{k_2}{\rightarrow} Cl{-}CH_3 + I^- \tag{6.14}$$

$$N_3^- + CH_3CH_2CH_2CH_2{-}Br \overset{k_2}{\rightarrow} CH_3CH_2CH_2CH_2{-}N_3 + Br^- \tag{6.15}$$

$$N_3^- + F{-}C_6H_4{-}NO_2 \longrightarrow [F,N_3 \text{ sigma complex } (-)]{-}NO_2 \longrightarrow N_3{-}C_6H_4{-}NO_2 + F^- \tag{6.16}$$

$$C_5H_{10}NH + F{-}C_6H_4{-}NO_2 \longrightarrow [\text{sigma complex}] \longrightarrow C_5H_{10}N{-}C_6H_4{-}NO_2 + HF \tag{6.17}$$

The most important practical consequence of the huge rate accelerations effected by dipolar aprotic (non-HBD) solvents is that many S_N2/S_NAr type reactions have become feasible under mild laboratory conditions, which was

Table 6.6 Relative rates of the S_N2 anion–molecule reactions 6.14 and 6.15 and of the S_NAr reactions 6.16 and 6.17 in protic and polar non-HBD solvents at 25 °C.

Solvents	$\log(k_2^{solvent}/k_2^{MeOH})$ for reaction			
	6.14	**6.15**	**6.16**	**6.17**
Protic solvents				
CH_3OH	0	0	0	0
H_20	0.05	0.8	—	—
$CH_3CONHCH_3$	—	0.9	—	—
$HCONH_2$	1.2	1.1	0.8	—
$HCONHCH_3$	1.7	—	1.1	—
Dipolar non-HBD solvents				
$(CH_2)_4SO_2$ (sulpholane)	—	2.6	4.5	—
CH_3NO_2	4.2	—	3.5	0.8
CH_3CN	4.6	3.7	3.9	0.9
CH_3SOCH_3	—	3.1	3.9	2.3
$HCON(CH_3)_2$	5.9	3.4	4.5	1.8
CH_3COCH_3	6.2	3.6	4.9	0.4
$CH_3CON(CH_3)_2$	6.4	3.9	5.0	1.7
N-methylpyrrolidinone	6.9	—	5.3	—
$[(CH_3)_2N]_3PO$	—	5.3	7.3	—

Source: Reichardt (2003: 249).

not the case as long as only common hydroxylic solvents were available. Another important element is the *reversal in nucleophilic reactivity* of small anions which can be effected by this solvent change and which also results from the contrasting solvation capabilities of the two kinds of solvent.

Not all nucleophilic substitution reactions exhibit an increase in rate on going to a dipolar aprotic solvent from a protic one. The reaction of 9-cyanofluorenyl anion (9-CFA) with benzyl chloride (6.18) is a case in point (Bowden and Cook 1968). The rate data plotted in Fig. 6.3 show that adding DMSO to an ethanolic medium causes a moderate retardation in rate for this reaction.

$$\text{9-CFA}^- \text{(CN)} + \phi CH_2Cl \xrightarrow{-HCl} \text{9-cyano-9-benzylfluorene (NC, }CH_2\phi) \tag{6.18}$$

This result is understandable, however, on the basis that the large charge-delocalized 9-CFA anion would be strongly stabilized by the polarizable DMSO (i.e. $\delta\Delta G_{tr}$ negative), and that stabilization of the transition state (increased charge delocalization) would not be sufficient to overcome this. According to Table 6.5 this system is denoted as subject to negative initial state control.

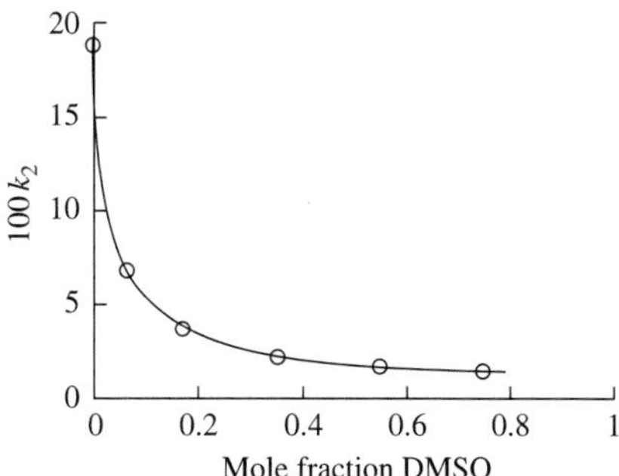

Fig. 6.3 Effect of added dimethyl sulphoxide on the rate coefficients, k_2, for the reaction of benzyl chloride with the 9-cyanofluorenyl anion in ethanolic media at 35.7 °C (Bowden and Cook 1968).

6.6 Proton transfer

Some of the largest rate enhancements due to solvent change have been observed for proton transfer processes. For example, the rate of the methoxide-catalysed racemization of 2-methyl-3-phenylpropionitrile, $PhCH_2CH(CH_3)CN$, is increased by a factor of 10^8 on going from methanol to DMSO (Cram *et al.* 1960). Racemization processes generally involve rate-determining proton abstraction followed by fast reaction with solvent to yield a planar carbanion. Fast protonation of this carbanion results in the formation of the two enantiomers in equal proportion.

Isotopic hydrogen exchange follows a similar mechanism, as shown in eqns 6.19 and 6.20 for the case of a protium-containing substrate undergoing exchange with deuterium-containing solvent.

$$\text{R–H} + \text{B}^- \xrightarrow{\text{slow}} \text{R}^- + \text{HB} \tag{6.19}$$

$$\text{R}^- + \text{D–OD} \xrightarrow{\text{fast}} \text{R–D} + \text{OD}^- \tag{6.20}$$

Isotopic hydrogen exchange has been investigated in a great variety of C–H containing substrates, including ketones, nitro compounds, nitriles, halides, sulphoxides, and aromatic hydrocarbons (Buncel *et al.* 1990; Buncel and Onyido 2002b; Buncel and Dust 2003). Generally, the isotopic exchange is facilitated when the negative charge in the carbanion can be delocalized through resonance, as shown for the carbanions derived from acetone, nitromethane, and toluene:

$$^{(-)}CH_2{-}C(=O){-}CH_3 \longleftrightarrow CH_2{=}C(O^{(-)}){-}CH_3$$

$$^{(-)}CH_2{-}\overset{+}{N}(=O){-}O^{(-)} \longleftrightarrow CH_2{=}\overset{+}{N}(O^{(-)}){-}O^{(-)}$$

$$C_6H_5{-}CH_2^{(-)} \longleftrightarrow {}^{(-)}C_6H_5{=}CH_2 \longleftrightarrow {}^{(-)}C_6H_5{=}CH_2$$

In the case of the trihalomethanes, which also undergo deuteroxide-ion catalysed H/D exchange (Symons and Clermont 1981), carbanion stabilization occurs through inductive electron withdrawal by the electronegative halogens. However, that this stabilizing effect is much smaller quantitatively than the conjugative effect is seen by the fact that chloroform, for example, is a weaker acid than $CH(NO_2)_3$ by ca. 20 pK units. Interestingly, for CF_3D/OH^-(0.01 M) at 49 °C, the rate of exchange increases by 6 orders of magnitude on going from water to 70 mol% DMSO (Symons *et al.* 1981).

An interesting case of isotopic hydrogen exchange is that in sulphoxides. Thus, deuterated dimethyl sulphoxide (widely used in NMR studies) can be prepared by allowing CH_3SOCH_3 (95% v/v) to equilibrate with OD^-/D_2O (5% v/v) at 100 °C several times and removing the exchanged HDO through distillation (Buncel *et al.* 1965). Apparently the basicity of hydroxide (deuteroxide) ion in DMSO-rich media is enhanced to a sufficient degree to allow facile proton abstraction to occur:

$$CH_3-\overset{\overset{\displaystyle O}{\|}}{S}-CH_3 \overset{OD^-}{\longrightarrow} HOD + CH_3-\overset{\overset{\displaystyle O}{\|}}{S}-CH_2^- \overset{DOD}{\longrightarrow} OD^- + CH_3-\overset{\overset{\displaystyle O}{\|}}{S}-CH_2D \qquad (6.21)$$

6.7 D_2–OH^- exchange

The isotopic exchange between molecular hydrogen (dihydrogen) and a hydroxylic solvent under base catalysis poses a rather intriguing mechanistic problem. The two main mechanisms that had been proposed (Symons and Buncel 1972) involve rate-determining proton transfer (eqns 6.22 and 6.23) or formation of an addition complex (eqns 6.24 and 6.25):

$$HO^- + D-D \overset{slow}{\longrightarrow} HOD + D^- \qquad (6.22)$$

$$D^- + HOD \overset{fast}{\longrightarrow} DH + OH^- \qquad (6.23)$$

$$HOH\cdots\underset{\underset{\displaystyle H}{|}}{O^-} + D-D \overset{slow}{\longrightarrow} HOH + [HO-D-D]^- \qquad (6.24)$$

$$[HO-D-D]^- + HOH \overset{fast}{\longrightarrow} HOD + DH + OH^-. \qquad (6.25)$$

The rate constant for exchange increases by ca. 10^4 on changing the medium composition from purely aqueous to 99.5 mole% DMSO at 65°. It is significant that this increase in rate is considerably less than observed in many other reactions for which the medium effect has been evaluated. Analysis in terms of thermodynamic and kinetic transfer functions gives information about the origin of the observed medium effect.

In Table 6.7 are presented the relevant data, derived from measured values of enthalpies of activation and available literature data for the enthalpies of transfer of hydrogen and hydroxide ion (Buncel and Symons 1976). It is apparent that there is a close parallel between the transfer enthalpies for the

Table 6.7 Enthalpies of Transfer (kcal mol^{-1}) for reactants and for the transition state of the D_2–OH^- exchange process in the DMSO–H_2O system

Mol % DMSO	$\delta\Delta H_{tr}^{D_2}$	$\delta\Delta H_{tr}^{OH^-}$	$\delta\Delta H_{tr}^{R}$	$\Delta H^{\ddagger}$	$\delta\Delta H^{\ddagger}$	$\delta\Delta H_{tr}^{T}$
0	0	0	0	24.0	0	0
10	1.2	−0.8	0.4	—	—	—
20.2	2.1	+3.7	5.8	21.2	−2.8	3.0
40.1	2.6	10.1	12.7	17.7	−6.3	6.4
59.0	2.2	13.8	16.0	16.4	−7.6	8.4
77.9	2.1	16.0	18.1	16.9	−7.1	11.0
87.5	2.1	16.9	19.0	16.8	−7.2	11.8
96.9	2.1	17.5	19.6	18.1	−5.9	13.7

Source: Buncel and Symons (1976).

reactants, $\delta\Delta H_{tr}^{R}$, and those for the transition state, $\delta\Delta H_{tr}^{T}$. There is some similarity with alkaline ester hydrolysis, in which case the transition state enthalpy transfer is also endothermic, though not nearly to the same degree as in the present system. Evidently, the destabilization of OH^- (noting that $\delta\Delta H_{tr}(OH^-)$ is the major component of $\delta\Delta H_{tr}^{R}$ in DMSO rich media) is largely retained in the transition state. The above results point to a rate-determining transition state with considerable charge localization on an electronegative atom. Any proposed mechanism must be in accord with this conclusion. The relative merits of various mechanisms in the light of this finding are considered in detail elsewhere (Buncel and Symons 1976).

7 Some other examples: Acidic, basic, chiral solvents; ionic liquids, green chemistry

7.1 Introduction

In this chapter we consider a few examples other than the dipolar-aprotic ones treated in Chapter 6. We have chosen solvents somewhat arbitrarily, perhaps because we have ourselves had occasion to use them, or because they exhibit particular properties that make them useful for special purposes. Sometimes mere inertness is all that is required, as in the case of solvents for chromyl chloride, mentioned in Chapter 2 (as it is a very active oxidant, it inflames immediately with diethyl ether, for instance). The properties required may include ability to dissolve a wide variety of substances, to sustain strong acidity, basicity, or oxidizing or reducing strength, a high or low boiling point, long liquid range, high or low dielectric constant, or low or (rarely) high viscosity. We continue by considering examples of solvents under four headings: acidic, basic, chiral, and ionic.

7.2 Acidic solvents

Acidic solvents may be used for their catalytic effect. Though many reactions are catalysed by protons present in ordinary solvents, including water, many other reactions require high levels of acidity to proceed at useful rates. These acidities are only attainable if the solvent is either so low in basicity that strong acids can be present in solution, or if it is itself moderately acidic. Acidity can sometimes be increased by the addition of a Lewis acid, as described in Chapter 3.

7.2.1 An acidic solvent; hydrogen fluoride

In spite of the well-known hazards associated with its use[1], and its chemical activity towards glass and many other materials, hydrogen fluoride has been used as a solvent since it was first prepared in the anhydrous state by Frémy (1856). Its boiling point, 19.5 °C, is not inconveniently low, but it is easily removed from a reaction mixture by distillation during workup. HF may be handled in copper, nickel, or Monel vessels, with minimal reaction, though

[1] **Caution**: HF gas is very irritating to mucous membranes. The liquid quickly penetrates the skin, and causes painful burns and necrosis. Rubber gloves, an efficient hood, and immediately available first-aid supplies are essential.

only Pt or Pt/Au alloy does not cause any contamination. Nowadays vessels lined with Teflon® or 'pctfe' (poly(chlorotrifluoroethene)) are used.

Pure HF is very strongly acidic ($H_o \approx -15$). Its autoprotolysis constant is comparable to that of water ($pK_{ip} = 13.7$). No common Brønsted acid is strong in HF (sulphuric, fluorosulphonic and perchloric acids are non-electrolytes), but through reactions with Lewis acids, such as

$$PF_5 + 2HF = H_2F^+ + PF_6^- \quad (7.1)$$

one can reach large negative values of H_o. Table 7.1 lists some substances exhibiting various degrees of acidity and basicity in HF (Dove and Clifford 1971: 174 and *passim*).

Spectroscopic measurements in HF are feasible using windows of quenched (glassy) pctfe (visible only), synthetic sapphire (UV-VIS), or fused silver chloride (IR). For NMR, samples can be sealed in Teflon or pctfe tubing, and encased in glass to permit spinning.

The solubilities of inorganic substances in HF resemble those in water, with some marked differences. Most elements are not dissolved, except the active metals, which react with evolution of hydrogen gas. Alkali-metal and some alkaline-earth salts are soluble, but many react liberating weaker acids in solution. Alkali-metal halides evolve the nearly insoluble hydrogen halides. Salts of transition metals are at most slightly soluble. Organic solubilities tend to be greater than those of the same substances in water. The most soluble substances are those carrying atoms or groups with EPD ability, such as O, N, S, or C=C.

7.2.2 Reactions in hydrogen fluoride

Chemistry in HF has been reviewed by Kilpatrick and Jones (1967) and by Dove and Clifford (1971), and particularly in its industrial aspects, by Smith (1994). The properties that make HF useful are its great acidity, its ability to dissolve a variety of organic and inorganic substances, and its resistance to oxidation. It is used in petroleum refining to carry out alkylations, such as the reaction of isobutane with light olefins to produce high-octane, branched C7 and C8 aliphatics that, being insoluble in HF, are readily separated from the solvent/catalyst. In the manufacture of detergents, similar alkylation reactions using HF as solvent and catalyst lead to a high proportion of the linear isomers, desirable because they are biodegradable (but see Section 7.5).

Table 7.1 Substances soluble in HF with acid/base properties

Acids	**Bases (not ordered)**
Appreciable (enhance acidity of HF): $SbF_5 > AsF_5 > NbF_5 > PF_5 > BF_3 > SnF_4$	MF: M = Li, Na, etc., NH_4, Ag, Tl, CH_3NH_3, $(CH_3)_4N$, $C_6H_5NH_3$, $(C_6H_5)_2NH_2$, etc.
Weak (react with F^-): $TiF_4 > VF_5 > MoF_5 > WF_6 > ReF_6 > GeF_4 > TeF_6 > SeF_4 > IF_4 > SiF_4$	MF_2: M = Ca, Sr, (Ba)

7.2.3 Electrochemistry in hydrogen fluoride

Studies of electrical conductivity in HF cited by Kilpatrick and Jones (1967) show mobilities of ions somewhat larger than in water. H^+ and F^- are more mobile than most ions (cf. H^+ and OH^- in water), suggesting a proton-jump mechanism. The concentration dependence of the conductance of all but a few electrolytes indicates considerable ion pairing, in spite of the high dielectric constant (83.6 at 0 °C). $Hg(CN)_2$, on the other hand, a non-electrolyte in most solvents, gives conducting solutions. Electrolysis of solutions of alkali fluorides in HF yields fluorine at the anode. HF has great affinity for water. If the solvent is not dry, the evolved fluorine is contaminated with OF_2, O_2, and ozone, but continued electrolysis is effective in removing water. Hydrocarbons and certain derivatives in HF solution are perfluorinated, at a less anodic potential than is required to liberate fluorine. For example n-hexane is converted to perfluoro-n-hexane according to reaction eqn (7.2):

$$C_6H_{14} + 42HF \rightarrow C_6F_{14} + 28H_2F^+ + 28e^- \qquad (7.2)$$

Except by electrolysis, HF is not easily oxidized. It should be a suitable solvent for strong oxidants. Dove and Clifford (1971) describe some such uses, but report experimental difficulties, involving reaction with reducing impurities, and attack by such very reactive solutes as rhenium heptafluoride and xenon hexafluoride on pctfe containers. Tables of electrochemical potentials are also presented.

7.3 Basic solvents

As was noted in Chapter 3, the presence of one or more unshared pairs of electrons confers both Brønsted and Lewis basicity, and nucleophilicity, on the molecule. This may be manifested as ability to solvate cations, to accept hydrogen bonds, to stabilize normal species or activated complexes by association with positive charges or electron-poor regions. Values of the empirical parameters that purport to measure basicity, β, B_j, DN, $pK_{BH+}(aq)$, etc., for different basic solvents accordingly differ even in their order of strengths. The hard/soft classification does not explain all the differences.

7.3.1 A basic solvent, ammonia

Organic solvents that are notably basic are almost all nitrogen-containing compounds; the simplest of these is ammonia. The physical properties of liquid ammonia are not very different from those of hydrogen fluoride. Its boiling point is lower (−33 °C), but not too inconveniently so.[2] It can be drawn from the cylinder as liquid or as gas, and is effectively dehydrated by distillation from the blue solution that it forms with sodium. Inorganic substances tend to be less soluble in ammonia than in water, especially salts of 1 : 2 or higher valence types. Reversals are observed where cations of salts

[2] **Caution**: Because it has a low boiling point and a relatively large heat of vaporization, and because it wets the skin, ammonia can quickly cause frostbite. It is poisonous by inhalation and requires adequate precautions.

form ammines, notably silver and lithium salts. Solubilities of organic substances in ammonia are often greater than in water. The most striking examples of substances with unusual solubilities are the alkali and alkaline-earth metals. The alkali metals dissolve without immediate reaction to form intensely blue (dilute) solutions. Water can also dissolve metals, as shown by the transient blue colour observed around the cathode when a concentrated NaOH solution is subjected to electrolysis at high current density. It is the basicity of ammonia that slows the decomposition. In the presence of ammonium salts, which are acids in ammonia, the reaction is rapid:

$$2NH_4^+ + 2Na \rightarrow 2NH_3 + H_2 + 2Na^+. \tag{7.3}$$

More concentrated solutions of alkali metals in ammonia (> 0.04 mole fraction of metal) have the appearance of liquid gold. The solubility of lithium in ammonia is large (>0.22 mole fraction), and most of the ammonia is bound in complexes: $Li(NH_3)_{4-x}$. The saturated solution consequently has a normal boiling point above room temperature. On standing the alkali metals react to form the amides with evolution of hydrogen, according to reaction (7.4):

$$2M + 2NH_3 \rightarrow 2M^+NH_2^- + H_2. \tag{7.4}$$

Li and Na react slowly, and the heavier members of the family more quickly, though none so vigorously that they cannot be effectively used. The alkaline earth metals are less soluble. These solutions appear to contain electrons, in the most dilute solutions solvated by ammonia. Stairs (1957) discusses a simple model for the solvated electron, trapped in a potential well created by polarization of the liquid. As the concentration is increased, the electrons become loosely bound to cations, and dimeric species begin to appear. To explain the concentration dependence of the activity, magnetic properties and conductance of these metal–ammonia solutions, Dye (1964) found it necessary to include a variety of species; the solvated metal ion M^+ and electron e^-, but also the 'atom' or ion pair M^+e^-, the triplets $e^-M^+e^-$ and $M^+e^-M^+$ and the quadruplet or 'dimer' $(M^+e^-)_2$. Feng and Kevan (1980) reviewed theoretical models. At the highest concentrations, the electrons can move among the cation sites as in a metal, and the electrical conductivity of the most concentrated solutions is comparable to that of mercury. The most useful chemical applications of these solutions are as reducing agents, as in the Birch reduction of aromatic hydrocarbons to alicyclics (e.g. see Pine 1987: 682, 936; Streitwieser *et al.* 1992: 634–636).

The acid–base range attainable in ammonia is large, corresponding to a pH range of 27 units or more, depending on temperature. H_- in the pure liquid at $-33\,°C$ is ≈ 22; in 0.1 M KNH_2 solution it is ≈ 35. Exchange reactions such as reaction (7.5)

$$NH_3 + D_2 \rightleftharpoons NH_2D + HD \tag{7.5}$$

are catalysed (e.g. by KNH_2), as are similar reactions of aliphatic amines. They are used for industrial production of deuterium (Buncel and Symons 1986).

Sodium amide is nearly insoluble in ammonia, but potassium amide, which is soluble, reacts with sodium and other insoluble metal amides to form soluble amido complexes, such as $Na(NH_2)_3^{2-}$ and $Al(NH_2)_4^-$. Amphoterism is thus much commoner in the ammonia system of compounds than in the water system (Audrieth and Kleinberg 1953: 81).

Ammonia is a levelling solvent for acids. All Brønsted acids except the weakest are converted to ammonium salts, which are acids in ammonia ('ammono-acids'). Owing to ammonia's rather low dielectric constant, ion pairing occurs, reducing the apparent acidity; the dissociation constants of ammono-acids as diverse as $(NH_4)_2S$ and NH_4ClO_4 range only from 9.8×10^{-4} to 5.4×10^{-3}. Nevertheless, solutions of ammonium salts in liquid ammonia will corrode some metals and dissolve many metal oxides. Divers's solution (a saturated solution of ammonium nitrate in ammonia), which is stable to near room temperature, acts on metals much as does aqueous nitric acid.

The solubility, properties in solution, and reactions of a large number of organic compounds in liquid ammonia were reviewed by Smith (1963), and of inorganic substances by Jander (1966) and by Lagowski and Moczygemba (1967) and again by Lagowski (1971) in the Symposium on Non-Aqueous Electrochemistry (Paris, July 1970). The last pay particular attention to electrochemical properties of both inorganic and organic substances in ammonia.

7.3.2 A basic solvent, pyridine

Pyridine is familiar to organic chemists as a reagent, as a reaction medium and as a component of chromatographic elution solvents. It is relatively easy to purify, though some related substances are not easily removed by fractional distillation. If such impurities are expected to interfere, fractional freezing is recommended. Its physical properties are convenient: a rather long liquid range (−40.7 to 115.5 °C), a moderate dielectric constant (12.5 at 20 °C), and a viscosity slightly less than that of water.[3]

Inorganic salts are more soluble in pyridine than its dielectric constant would lead one to expect, presumably owing to solvation of the cation through the nitrogen (the unshared pair of electrons occupies a non-bonding σ type orbital, not part of the π system of the ring, so is available). Solubility is favoured by high charge-to-size ratio for the cation, but the reverse for the anion. Pyridine is classified as a borderline base (Chapter 3). In a conductivity study (Hantzsch and Caldwell 1908) it appeared to differentiate the (hard) strong acids; $HI > HNO_3 > HBr > HCl$, but it is not clear that ion pairing was absent. The apparent dissociation constant of perchloric acid in pyridine is 7.55×10^{-4}. A scheme such as the following (at least) should be assumed:

$$Py + HClO_4 \rightarrow [PyH^+ClO_4^-] \rightleftharpoons PyH^+ + ClO_4^- \tag{7.6}$$

[3] **Caution**: Pyridine is reported to cause gastrointestinal upset and central nervous system depression at high levels of exposure, and to depress sexual activity in men. It can be absorbed through the skin; rubber gloves are recommended. The vapour pressure of pyridine at 20 °C is 14.5 Torr. The resulting concentration in air would be four times the maximum allowed concentration, $15\,mg\,m^{-3}$, but it is detectable by its strong odour at much lower concentrations.

Pyridine can sustain a large range of oxidation and reduction potential, depending on the electrolyte present and the electrode material, of more than 3 V. Metals as active as Li, K, and Ba have been successfully electrodeposited from strictly anhydrous pyridine solutions of suitable salts onto Pt or Fe electrodes, or into mercury, though Nigretto and Jozefowicz (1978) in a review of the uses of pyridine as a solvent in analytical, and especially electroanalytical, chemistry, list a number of reagents that reduce pyridine. Pyridine is oxidized by persulphate, but CrO_3 dissolves to form Sarett's reagent (Poos *et al.* 1953; House 1972), a strong but selective oxidant, used to oxidize alcohols to carbonyl compounds (but now superseded by pyridinium chlorochromate). Scriven *et al.* (1994) describe pyridine as the solvent of choice for acylations (but see Section 7.5), and as excellent for dehydrochlorinations, owing to its ability to act as a scavenger for acid.

A well-known use of pyridine as a cosolvent is in the Karl Fischer titration of water in organic solvents, described in Vogel *et al.* (1978). A solution of iodine and sulphur dioxide in pyridine/methanol or pyridine/cellosolve is fairly stable in the absence of water, but when it is added to a sample of a solvent containing water, reaction (7.7) occurs quantitatively.

$$3C_5H_5N + I_2 + SO_2 + ROH + H_2O \rightarrow 2C_5H_5NH^+I^- + C_5H_5H^+ROSO_3^- \tag{7.7}$$

(R is CH_3 or $CH_3OCH_2CH_2$.) The endpoint is detected by the persistence of the brown colour of iodine, or electrometrically.

7.4 Chiral solvents

The use of chiral, nonracemic solvents in methods of separation of enantiomers has been reviewed by Eliel *et al.* (1994). They conclude that separation based on solubility difference between enantiomers in a chiral solvent is unlikely to lead to a practical method of separation. On the other hand, partition between immiscible solvents, one of which is chiral, can lead to separation. In most cases, the degree of separation in one equilibration is very small, but it has been shown to permit at least partial resolution of racemic mixtures by multiple extraction (Bowman *et al.* 1968). There seems to be a requirement of a definite complex between the molecules being separated and the chiral solvent molecules, in which at least two hydrogen bonds are present, to give a degree of conformational rigidity to the complex. Merely surrounding the enantiomers with chiral solvent molecules is insufficient, as is the formation of a single H-bond. Striking success was achieved by Cram and coworkers (Kyba *et al.* 1973; Cram and Cram 1978), by the use of synthetic 'host' molecules (example illustrated) having a rigid structure and several H-bonding sites directed to the interior of a cavity, into which one 'guest' enantiomer with corresponding outward-directed H-bond sites fitted, and the other enantiomer did not. Peacock and Cram (1976) reported the resolution of phenylglycine with 96% enantiomer purity in one extraction.

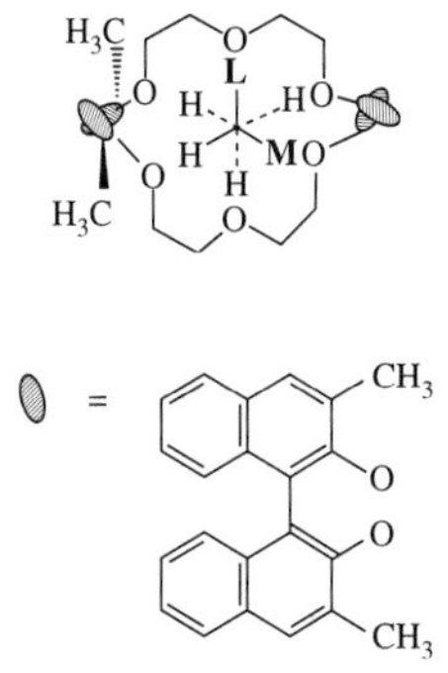

A reaction that in an achiral solvent would produce a racemic product, when carried out in a chiral solvent may result in the predominance of one of the enantiomers. This may result either from differential solvation of the reactants or of the transition state. Bosnich (1967) has shown that a symmetric solute in a chiral solvent may exhibit induced asymmetry, which can influence the ratio of enantiomers formed. Where the effect is through the transition state, one would expect the effect to be greatest if the solvent were involved directly, especially (in view of the foregoing) if two or more H-bonds are involved. If A and B are non-chiral molecules, but their adduct AB exists as enantiomers AB_+ and AB_-, the activated complexes in the reactions;

$$A + B + S_+ \rightarrow [SAB_{++}]^\ddagger \rightarrow S_+ + AB_+ \tag{7.8}$$

and

$$A + B + S_+ \rightarrow [SAB_{+-}]^\ddagger \rightarrow S_+ + AB_- \tag{7.9}$$

are diastereomers, so the reactions would be expected to proceed at different rates, and yield more of one product than the other. If the reactions are reversible, and proceed to equilibrium, the racemic mixture will result, unless the solvent forms diastereoisomeric complexes with the products, in which case some enantioselectivity may still be obtained.

Rau (1983), in a review of asymmetric photochemistry in solution, briefly describes some effects of chiral solvents in photochemical synthesis. He mentions optical purities of products ranging from less than 1 to 23.5%. Reichardt (2003: 69) cites similar figures for a number of reactions, concluding that, while chiral solvents can induce asymmetry (or enhance the purity of products of reaction between chiral reactants) the effects are usually rather small.

7.5 Ionic liquids

Until fairly recently 'ionic liquids' meant 'fused salts', and referred to substances with melting points well above room temperature (e.g. see Bloom and Hastie 1965; Kerridge 1978). By using binary or ternary eutectics, lower working temperatures were attainable (e.g. the ternary eutectic of lithium, sodium and potassium nitrates, 125 °C), but these systems attracted little interest from organic chemists. Inorganic chemists, however, have found molten salts or salt mixtures useful for preparation of species otherwise unavailable. For instance, in liquid LiCl, the concentration of Cl^- is high enough to cause the formation of species such as $FeCl_4^{2-}$ or $CrCl_6^{3-}$. In the curious room-temperature liquid, triethylammonium chlorocuprate(I), even $FeCl_6^{4-}$ can exist. The alkali metal halides dissolve notable amounts of their parent metals, giving solutions that resemble the metal–ammonia solutions in their optical, electrical, and magnetic properties. Some other metals may dissolve in their halide melts with reaction. For instance, while a solution of cadmium in fused cadmium chloride gives back the metal on cooling and solidifying, addition of $AlCl_3$ causes the precipitation of $Cd_2^{2+}[AlCl_4^-]_2$, containing the dimeric cadmium(I) ion, analogous to mercurous ion, Hg_2^{2+}.

Many studies have been related to extractive metallurgy or to investigation of reactions responsible for the formation of minerals, including ores. The chemistry of molten oxide mixtures led to the formulation by Lux (1939) of the acid-base theory, refined by Flood *et al.* (1952), based on oxide ion as the species transferred, analogous to the proton in the Brønsted–Lowry view, though opposite in sense, for an oxide donor is a base, and an oxide receptor is an acid. Many authors put 'acid' and 'base' in quotation marks, or use such phrases as 'acid analogue' in discussing the chemistry of such melts. This is quite unnecessary, however, as the analogy with protonic acidity is perfect, *mutatis mutandis*. Highly silicic lavas, slags, etc., are acidic, and rocks with less than about 50% SiO_2 can be called 'basic' without apology, as demonstrated by measurements of pO using suitable electrodes (El Hosary *et al.* 1981).

Ethylammonium nitrate, m.p. 14 °C, was described by Walden in 1914, but did not attract wide attention. The discovery of the 2 : 1 aluminium chloride/ethylpyridinium bromide combination, which is also liquid at room temperature (Hurley and Weir 1951) began a new chapter, introducing a family of systems involving $AlCl_3$ and halides of large organic cations. Acid–base properties of $AlCl_3$/MCl systems (M^+alkali metal cation) had previously been studied, using an Al wire electrode to measure Cl^- activity. The neutral species (in the acid/base sense) is the $AlCl_4^-$ anion. The reaction (called by Chum and Osteryoung (1981) *autosolvolysis*):

$$2AlCl_4^- \rightleftharpoons Al_2Cl_7^- + Cl^- \quad K_m \qquad (7.10)$$

is analogous to autoprotolysis in water. $Al_2Cl_7^-$ is the acid species, and Cl^- the basic species; an unusual pair, as both are anions.

The equilibrium constant for this reaction is strongly temperature dependent and somewhat influenced by the nature of the cation. Chum and Osteryoung report values of K_m of 1.06×10^{-7} in $AlCl_3$/NaCl and $< 1.19 \times 10^{-8}$ in $AlCl_3$/N-n-butylpyridinium chloride (RCl) at 175 °C, and 3.8×10^{-13} in the latter system at 30 °C. They show what are effectively titration curves, showing a potential jump of more than −0.8 V as the mole ratio of $AlCl_3$/RCl passes from just below to just above unity, corresponding to a change in pCl at 30 °C of nearly 14 units, fully comparable to the pH jump in the titration of aqueous 1 M NaOH with strong acid.

Other anions than the very water-sensitive and reactive haloaluminates have been added to the repertoire, including nitrate, phosphate, hexafluorophosphate, hexafluoroantimonate, tetrafluoroborate, trifluoromethylsulphonate ('triflate'), and bis(trifluoromethanesulphonyl)imide, $(CF_3SO_2)_2N^-$. Examples of cations that are used include mono- to tetra-alkylammonium and phosphonium, 1-alkyl-3-methylimidazolium, **1**, N-alkylpyridinium, **2**, and the ions **3**, **4**, **5**, and **6** containing more than one quaternary nitrogen atom. A wide variety of properties is thus available, including water solubility from near zero to total miscibility, and selective solubility of different classes of organic substances. Plešek and Heřmánek (1968), in discussing the uses of

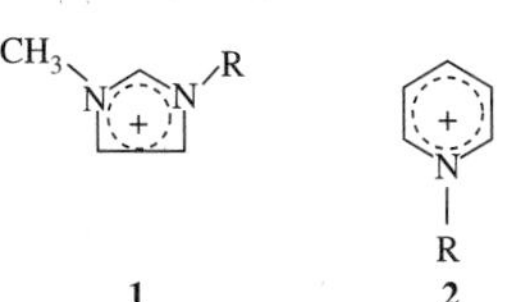

sodium hydride, comment on its 'complete insolubility in the usual solvents', though it is soluble in molten sodium hydroxide. One wonders if there are any room-temperature ionic liquids that would dissolve sodium hydride without reacting with the exceedingly strongly basic hydride ion (Buncel and Menon 1977b).

Room-temperature ionic liquids are attracting a great deal of industrial interest (Guterman 1999; Carmichael 2000). They are versatile as solvents or non-solvents for organic substances, and some exhibit strongly temperature dependent water solubility. They are non-volatile. Some, notably those containing haloaluminate anions, have widely variable Lewis and Brønsted acidity (into the superacid range). Others, such as those with tetrafluoroborate or hexafluorophosphate anions, lack strong acidic or basic properties. These unreactive examples have the ability to dissolve metal-containing catalysts without deactivating them, permitting recovery and recycling of catalysts that would otherwise have been lost to waste. By adjustment of the temperature, certain reaction mixtures can be made homogeneous for reaction, and then cooled to cause phase separation for recovery of products; or the reaction mixture can be maintained as a two-phase system while reaction proceeds.

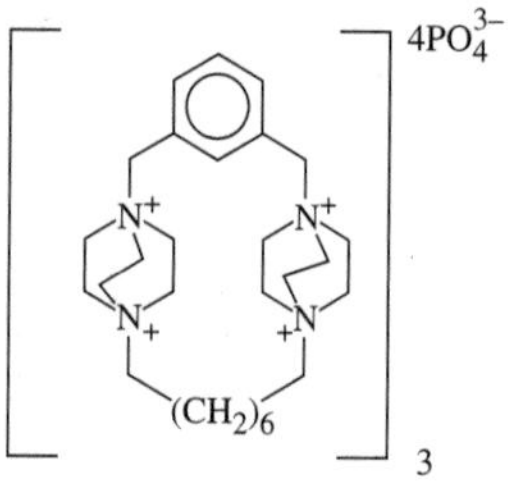

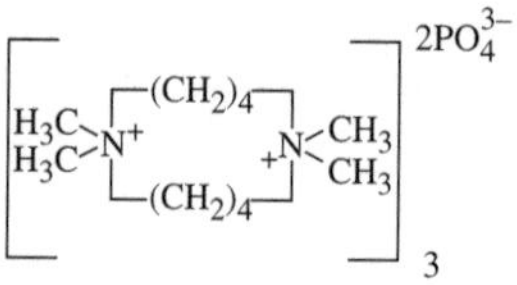

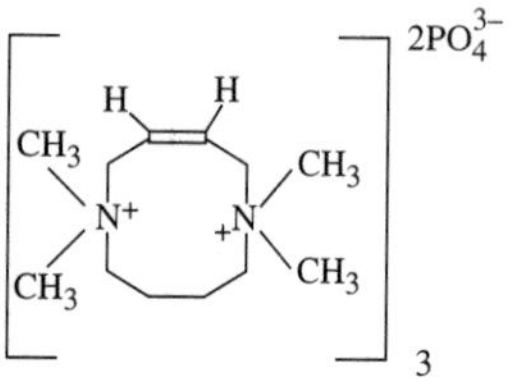

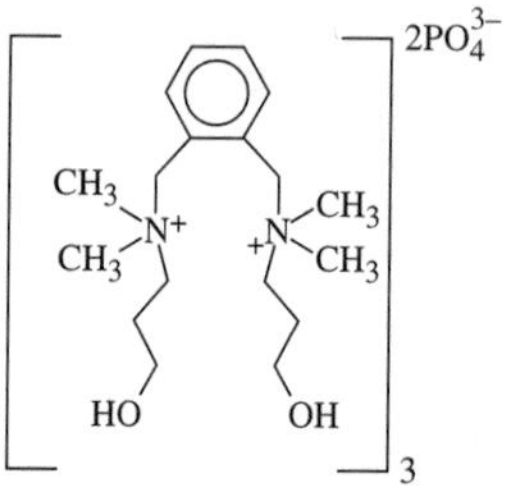

7.5.1 Reactions in ionic liquids; green chemistry

Ionic liquids, owing to their stability and non-volatility, contribute to 'green' chemistry. In many applications they are easily recovered and recycled, and volatile organic substances can be recovered from them by distillation under mild conditions *in vacuo*. For instance, the ability of 1-butyl-3-methylimidazolium hexafluorophosphate to dissolve both the organic reagent benzyl chloride and the salt potassium cyanide has been exploited (Wheeler *et al.* 2001) in a successful preparation of phenylacetonitrile, with minimal environmental impact.

A number of reaction types may be advantageously carried out in room-temperature ionic liquids. They include the following:

- *Friedel–Crafts.* The variable acidity afforded by chloroaluminate ionic liquids enables improved control of the course of Friedel–Crafts acylations and alkylations over traditional solvents. Boon *et al.* (1986) showed that the rate of acetylation of benzene (reaction 7.11) by acetyl chloride in chloroaluminate liquids was directly proportional to the concentration of $Al_2Cl_7^-$. That the steric course of reaction may be controlled is illustrated by the predominance of α- over β-acetylation of naphthalene in 1-ethyl-3-methylimidazolium chloride/$AlCl_3$ (Adams *et al.* 1998); the reverse is true in nitrobenzene as solvent. Alkylation of benzene with 1-dodecene, using $[(CH_3)_3NH]^+ \cdot Al_2Cl_7^-$ as catalyst gives a better yield of linear dodecylbenzene for detergent manufacture, with less loss of catalyst, than the similar process using liquid HF (P. de Jonge, cited by Carmichael 2000). A variant method uses a neutral ionic liquid, and scandium(III) triflate as an immobilized Lewis acid catalyst (Song *et al.* 2000).

$$C_6H_6 + CH_3COCl \xrightarrow{Al_2Cl_7^-} C_6H_5COCH_3 + HCl \qquad (7.11)$$

- *Diels–Alder.* The activity as dienophile of unsaturated esters, or aldehydes such as crotonaldehyde (2-butenal), towards cyclopentadiene in dichloromethane was enhanced by the addition of 0.2 equivalent of 1,3-diethylimidazolium bromide or acetate (Howarth *et al.* 1997). Diels–Alder reactions have also been carried out in ionic liquids as the sole solvents, with or without the addition of Lewis acid catalysts (Song *et al.* 2001). The yields were generally satisfactory, and exo/endo selectivity moderately strong (in the region 4.3–4.9 to 1), comparable to ratios found in polar solvents. Notably, for reaction (7.12) in 1-ethyl-3-methylimidiazolium chloride/aluminium chloride solvents, as the solvent composition was changed from the basic to the acidic region, the selectivity changed from 4.88 : 1 to 19 : 1, with some loss of yield. Kumar (2001) comments that it is too soon to assess or explain the effects of these ionic media on Diels–Alder reactions, and that more kinetic data are required. He refers to the review by Welton (1999) for a discussion of the properties of ionic liquids that are relevant to their usefulness in this context, and in organic chemistry generally.

$$C_5H_6 + CH_2{=}CHCO_2CH_3 \xrightarrow{\sim R.T.} \text{(endo)} + \text{(exo)} \qquad (7.12)$$

- *Hydrogenation and hydroformylatiom.* Platinum group metals, in the form of low-oxidation-state complexes or cluster compounds, dissolved in ionic liquids such as 1-butyl-3-methylimidazolium tetrafluoroborate have been used as hydrogenation catalysts for both aliphatic (Suarez *et al.* 1997) and aromatic (Dyson *et al.* 1999) substrates. The reactions proceed heterogeneously, and the product is easily separated. The catalyst remains in the solvent for re-use. This method has been used successfully to hydrogenate acrylonitrile-butadiene rubber, rendering it less susceptible to thermal and oxidative degradation (Suarez *et al.* 1997). Hydroformylation reactions (e.g. reaction (7.13)) in ionic liquids afford yields similar to those obtained in dichloromethane as solvent, but separation of the product is easier, and loss of a volatile organic substance to the environment is minimized.

$$RHC{=}CH_2 + CO + H_2 \longrightarrow RCH_2CH_2CHO + RCH(CH_3)CHO \qquad (7.13)$$

- *Oligomerization*. Dimerization or oligomerization of low molecular weight alkenes is regularly carried out industrially, with the aim of preparing linear 1-alkenes for synthetic use. In these applications, the presence of strong Lewis acids is undesirable, so neutral ionic liquids are used, exemplified by [bmim]PF_6. ($[bmim]^+$ is 1-butyl-3-methylimidazolium.) This has been successful in the manufacture of hexenes from ethylene using a nickel complex as catalyst (Wasserscheid *et al.* 2001), with conversion of 95%, the product being 92% 1-hexene. An earlier study by Chauvin *et al.* (1990) on the dimerization of propene demonstrated the pitfalls of using chloroaluminate liquids: in the acid region, cationic side reactions led to too great molecular weights and high viscosity, while in the basic region, chloride displaced ligands from the nickel catalyst, deactivating it. It was shown that the system may be buffered by adding an insoluble solid alkali-metal chloride to the liquid on the acidic side (Ellis *et al.* 1999), which converts excess $Al_2Cl_7^-$ into neutral $AlCl_4^-$ and at the same time provides alkali cations to precipitate chloride, preventing an excursion into the basic region. Alternatively, replacing $AlCl_3$ with $EtAlCl_2$, which is less acidic, led to successful production of the desired dimers.

An avenue that is now being explored involves the use of 'task specific' ionic liquids, in which one of the ions is furnished with a functional group for a particular purpose. An example shown is reaction (7.14), developed by Davis and co-workers (Bates *et al.* 2002). The amino group binds carbon dioxide as

$$[\text{Bu-im}^+\text{-(CH}_2)_3\text{NH}_2]\,BF_4^- + CO_2 \underset{\text{R.T.}}{\overset{80\text{–}100\,^\circ\text{C (vacuum)}}{\rightleftharpoons}} \text{Bu-im}^+\text{-(CH}_2)_3\text{NH-C(=O)O}^- \qquad (7.14)$$

a carbamate at room temperature in a reaction that is reversed at a higher temperature. It is proposed as a substitute for aqueous amines now used to scrub carbon dioxide from natural gas. The new process avoids problems with loss of the amine and of water. A problem with viscosity was encountered, which may perhaps be overcome by changing the butyl side chain, or by suitable dilution.

This is an example of a type of process that should become important in many industries, for extraction of toxic or valuable substances from gaseous or liquid waste streams. Ionic liquids can be prepared that are essentially immiscible with water. A suitable ionic liquid denser than water coming down a packed tower could extract many organic or inorganic solutes from an ascending aqueous stream. Losses of the ionic solvent to the water are expected be very low, and the solvent and the extracted substances should be recoverable by moderate heating under reduced pressure.

8 Concluding observations

8.1 Choosing a solvent

Up to this point, we have been considering a number of aspects of solvents and their effects, partly from a theoretical point of view, or as empirical observations. The reader may well say, 'What now?' and expect some practical advice. What this advice will be must depend on the purpose of the procedure being undertaken.

Carlson *et al.* (1985) describe an approach to a formal method of choosing the most suitable solvent, based on a Principal Component Analysis (see Chapter 4) of the properties of a set of solvents of diverse types. Using 82 solvents and eight descriptors (m.p., b.p., relative permittivity ϵ_r, dipole moment μ, refractive index n_D^{20}, $E_T(30)$, density ρ, and log $K_{o/w}$), they show that 51% of the variance of the solvent properties can be represented by two principal vectors (components), $\boldsymbol{t}_1$ and $\boldsymbol{t}_2$, and that a third does not add significance (it may be observed that the descriptors chosen are all of the non-specific type, with the exception of $E_T(30)$). They then propose an experimental design based upon a systematic exploration of the surface defined by the vectors $\boldsymbol{t}_1$ and $\boldsymbol{t}_2$. The point of this formal procedure is to avoid premature choice of a single solvent type based upon an assumed mechanism, which, if it should prove unproductive, might lead the investigator to give up. They do not claim to have described a definitive procedure, but rather they point in a direction in which future work might proceed.

In the absence of a suitable formal method of choice, the choice must be guided by the information available, and the purpose of the work. Is the solvent to be a reaction medium, either for synthesis or for a mechanistic study? Practically, it must have a reasonable liquid range, be reasonably easy to purify, be separable from reaction products (by evaporation or by their low solubility), and not react destructively with the reactants or hoped-for products. Consideration of yield as described in Chapter 2 may or may not be important, though the effects of changing solvents may be decisive in assigning a mechanism. Certain reactions may require a two-phase system, if one reactant is ionic and the other relatively non-polar. For instance, in the emulsion polymerization method of making certain synthetic rubbers (Billmeyer 1971), a mixture of monomers (styrene and butadiene in the classic case, 'GR–S', important during the Second World War), neat, is dispersed with the help of soap in an aqueous solution of the initiator, a source of radicals. Initiators used have included potassium persulfate and a reducing agent, and Fenton's reagent, which contains hydrogen peroxide and ferrous sulphate. The soap forms micelles containing monomer. HO radicals diffuse across the phase boundary and initiate radical polymerization within the micelle. Concentrations can be adjusted so that on average only one radical at a time enters each micelle, so each contains a single growing polymer.

A simplified scheme of reaction is as follows:

$$\mathrm{HOOH(aq) + Fe^{2+}(aq) \rightarrow HO^{\bullet}(aq) + OH^{-}(aq) + Fe^{3+}(aq)} \quad \text{(initiation)}$$

$$\mathrm{OH^{\bullet}(aq) \rightarrow OH^{\bullet}(org)} \quad \text{(phase transfer)}$$

$$\left.\begin{array}{l} \mathrm{OH^{\bullet} + M \rightarrow HOM^{\bullet}} \\ \quad\vdots \\ \quad\vdots \\ \mathrm{HOM_i^{\bullet} + M \rightarrow HOM_{i+1}^{\bullet}(= R_{i+1}^{\bullet})} \end{array}\right] \quad \text{(propagation)}$$

$$\mathrm{R_i^{\bullet} + R_j^{\bullet} \rightarrow P_{i+j}} \quad \text{(termination)}$$

M may be either of the two monomers; $HOM_i^{\bullet}$ or $R_i^{\bullet}$ represents an active polymeric radical containing *i* monomer units; P_i represents a 'dead' polymer. Chain-transfer reactions, in which a polymer radical abstracts a H atom from a monomer (starting a new chain) or from the middle of another polymer molecule (resulting in a branched structure), and other modes of termination are possible.

Some such two-phase reactions require a phase-transfer catalyst, usually a large organic ion, such as tetrabutylammonium, to carry an ionic reagent into the non-polar phase.

Solvents for recrystallization must satisfy a fairly rigid set of criteria. The substance to be purified must be much more soluble at high than at low temperature. Expected impurities should be either very soluble, or not at all. The solvent must be easily separated from the crystals by evaporation, or soluble in a second solvent in which the crystals are not, so it may be washed off.

Solvents for chromatography, for electrochemistry, or for titrimetric analysis of weak acids or bases all must satisfy appropriate criteria. Enzymatic reactions have special requirements, some of which are surprising to the uninitiated, who may assume that biochemistry is aqueous chemistry. Yamane *et al.* (2001) point out that many enzymes exhibit much greater thermal stability in dry organic solvents than in water, that in many cases specificity is enhanced, that stereospecificity is sometimes altered, and may even be reversed. Freemantle (2000) cites industrial research into the use of ionic liquids as media for enzyme-catalysed reactions.

8.2 Envoi

Much of what has been introduced in this book, and a great deal that is not, is described in Wypych's (2001) 'Handbook of Solvents'. It contains information on uses of solvents in a wide variety of industries, toxicology and environmental effects of solvents, their safe use and disposal, recommendations for substitution of solvents by safer substances and processes, protection of workers from solvent exposure, detection and control of solvent residues in products, and regulations in effect in the United States, Canada, and the European Economic Community. It concludes with a section on environmental contamination cleanup.

These and other considerations are admirably discussed, and tables of relevant data provided, in the appendix to Reichardt's book, 'Solvents and Solvent Effects in Organic Chemistry' (2003: 471), to which reference has been repeatedly made in these chapters. This appendix also provides information on the toxicity of solvents and their safe handling, as well as methods of purification appropriate to various classes. We could not do better than to conclude by reiterating our recommendation of this book and especially its appendix.

Appendix

Table A.1 Properties of selected solvents (at 25 °C except as noted)

Solvent	mp(°C)	bp(°C)	ρ(g cm^{-3})	η(cP)	ϵ_r	n_D^{20}	Hazard (Fp)[a]	Disposal[b]
Fluorocarbons								
f-n-hexane[c]	−4	55–60	1.669			1.2515		A
f-benzene	4	82	1.612			1.3769	(10)	D
Hydrocarbons								
n-hexane	−95	69	0.659	0.294	1.90	1.3749	(−23)	D
Cyclohexane	6.5	81	0.779	0.895	2.023^{20}	1.4260	(−18)	D
Benzene	5.5	80	0.874	0.601	2.27	1.5010	CARC?, SK (−11)	D
Toluene	−93	111	0.867	0.550	2.38	1.4968	(4)	D
Halogenated hydrocarbons								
1,1,2-trichloro-*f*-ethane	−35	48	1.575		2.48	1.3578		A
Dichloromethane	−97	40	1.325	0.425	8.9	1.4240	POI, IRR	B
Trichloromethane	−63	61	1.492	0.542	4.7	1.4460	CARC?	B
Tetrachloromethane	−23	77	1.594	0.880	2.23	1.4595	POI, CARC?	B
1,2-dichloroethane	−35	83	1.256	0.39^{30}	10.36	1.4438	CARC? (15)	D
Chlorobenzene	−45	132	1.107	0.80^{20}	5.62	1.5241	POI	C
Hydroxy-compounds								
Water	0.0	100.0	0.9971	0.891	78.30	1.3330		
Water, crit. (supercrit. properties widely variable)	—	$t_c = 374$ (220 Bar)	0.4			1.27(est)	(high pressure, temperature)	
Methanol	−98	64.8	0.7866	0.553	32.63	1.3284	POI (11)	D
Ethanol	−114.5	78.3	0.7851	1.06	24.3	1.3610	(8)	D
1-propanol	−126	97.2	0.7998	2.27	20.1	1.3854	IRR (15)	D
2-propanol	−89	82.3	0.7854^{20}	1.77^{30}	18.3	1.3770	(12)	D
2-methyl-2- propanol	25.5	83	0.786	2.07	10.9^{30}	1.3860	(4)	D
1,2-ethanediol	−13	198	1.113	16.2		1.4310	POI (110)	A
Ethers								
Diethyl ether	−113	34.6	0.706	0.222	4.22	1.3506	(−40)	D
Di-n-butyl ether	−98	143	0.764			1.3988	(25)	D
Tetrahydrofuran	−108	67	0.886		7.39	1.4070	(−17)	D
1,4-dioxane	11.8	102	1.034		2.21	1.4215	CARC? (12)	D
1,2-dimethoxyethane	−74	56	0.842			1.3923	(0)	C
Ketones								
Acetone	−94	32	0.791	0.316	20.7	1.3585	(−17)	D
2-butanone	−87	80	0.805		18.5^{20}	1.3788	(−3)	D
Esters								
Methyl formate	−99	32	0.974		8.5^{20}	1.3425	(−27)	D
Ethyl acetate	−84	77	0.902	0.441	6.02	1.3720	(−3)	D
γ-butyrolactone	−45	205	1.120			1.4365	(98)	C
Propylene carbonate	−55	240	1.189	2.53	65.1	1.4210	(132)	A
Nitrogenous compounds								
Ammonia	−77.7	−33.4	0.65^{-10}	0.25^{-33}	16.9		COR	N
2-propylamine	−101	34	0.694	0.724	5.5^{20}	1.3746	COR, POI (−30 est)	D
Aniline	−6	184	1.022	3.71	6.89^{20}	1.5863	POI, SK, CARC? (70)	C

Table A.1 (*Contd*)

Solvent	mp(°C)	bp(°C)	ρ(g cm)$^{-3}$	η(cP)	ϵ_r	n_D^{20}	Hazard (Fp)[a]	Disposal[b]
Acetonitrile	−48	82	0.786	0.345	36.2	1.3440	LACH (5)	D
Benzonitrile	−13	188	1.010	1.24		1.5280	IRR (71)	C
Pyridine	−42	115	0.978	0.954^{20}	12.3	1.5102	POI (20)	C
Formamide	8.4	210 d	1.129	3.76^{20}	109.5		TER, IRR (??)	A
N,N-dimethylformamide	−61	153	0.944	0.796	36.7	1.4305	IRR (57)	C
Nitromethane	−29	101	1.127	0.608	38.6^{20}	1.3820	(35)	D
Nitrobenzene	6	211	1.196	2.03^{20}	34.6	1.5513	POI, SK (87)	C
Hexamethylphosphoramide	7	232	1.030	3.5^{60}	30^{20}	1.4579	POI, CARC? (105)	A
Tetramethylurea	−1	177	0.971			1.4506	(65)	C
Carboxylic acids								
Formic acid	8.4	101	1.220	1.80^{20}	57.9^{20}	1.3704	POI, COR (68)	C
Acetic acid	16	117	1.059	1.16	6.19	1.3715	COR, SK (40)	C
Sulphur compounds								
Dimethylsulphoxide	18.4	189	1.101	1.98	47.6^{23}	1.4787	SK (95)	A
Sulpholane	27	285	1.261	9.87^{30}	44.0^{30}	1.4840	(165)	A
Sulphuric acid	10.5	330 d	1.840	24.5	100		COR (–)	N
Sulphur dioxide	−75.5	−10	1.434^{-10}	0.428^{-10}	12.3^{22}	1.410	POI, IRR (–)	K
Carbon disulphide	−112	46	1.266	0.363	2.64^{20}	1.6270	POI (−33)	D
Other acids								
Phosphoric acid	42	213 d	1.834^{18}	178	~61		COR (–)	N
Hydrogen fluoride	−83	19.4	0.987	$0.24^{6.25}$	60^{19}		POI, COR, SK (–)	F
Hydrogen cyanide	−14	26	0.699^{20}	0.20^{20}	106.8	1.2675^{10}	POI (??)	
Ionic liquids								
LiCl/NaCl 25 : 75 mol/mol	551						(–)	
$LiCl/AlCl_3$ 35 : 65	80						(–)	
$[1\text{-ethyl-mim}]^+ [AlCl_4]^-$[d]	7		1.266	$\sim 18^{25}$			(–)	
$[1\text{-ethyl-mim}]^+ [Al_2Cl_7]^-$	ca. −20			9.5			(–)	
$[1\text{-ethyl-mim}]^+ [CH_3COO]^-$				162^{20}			(–)	
$[1\text{-butyl-mim}]^+ [CF_3SO_3]^-$[e]	16			90^{20}				
Other inorganics								
Carbon dioxide (crit.)		$t_c = 31.1$ (74 bar)	0.460				(high pressure) (–)	(vent)
Arsenic trifluoride	−8.5	63	2.67		5.7		POI (–)	L
Thionyl chloride	−105	79	1.631		9.0	1.5140	LACH (–)	N
Phosphorus oxychloride	1.25	105.8	1.645		13.9^{22}	1.460^{25}	CORR (–)	N

Notes

[a] POI = poisonous by inhalation or ingestion; IRR = irritating, especially to mucous membranes and eyes; SK = readily absorbed through the skin; LACH = lachrymator; CARC(?) = carcinogenic (suspected); TER = teratogenic. CORR = Corrosive; (Fp) = Flash point (where applicable).

[b] Waste-disposal methods adapted from those recommended in the annual catalogue of Aldrich Chemical Co. Ltd., 940 W. Saint Paul Ave., Milwaukee, Wis. 53233, USA:

A. Mix with combustible diluent, incinerate. 'Incinerate' here means burn in a chemical incinerator equipped with afterburner and scrubber.
B. Ignite in the presence of sodium carbonate and slaked lime.
C. Incinerate.
D. Incinerate, with precautions due to its high inflammability.
F. Add to excess of water, neutralize with sodium carbonate. Add sufficient calcium chloride to precipitate fluoride and carbonate. Dispose of solids to secure landfill.
K. Consult supplier.
L. Add to large excess water, precipitate as sulphide, neutralize, oxidize excess sulphide with hypochlorite. Solids to secure landfill, liquid to drain.
N. Add to large excess of water, neutralize with slaked lime. Separate solids for disposal to secure landfill. Flush aqueous solution to drain.

[c] Perfluorinated substance designated by *f*.

[d] 'mim' is 3-methylimidazolium cation. (Fannon *et al.* 1984)

[e] (Wasserscheid and Keim 2000).

Table A.2a Solvent property parameters: symmetric properties

Symbol	Measure of (Basis)	Author, Ref.
Part 1: Solvatochromic parameters		
$A(^{14}N)$	Polarity (ESR hyperfine splitting)	Knauer and Napier (1974)
E_K	Polarity (energy of $d \rightarrow \pi^*$ transition in a Mo complex)	Walther (1974)
E^*_{MLCT}	Similar, based on a W complex	Manuta and Lees (1986)
$E_T(30)$	Polarity (CT abs freq of a dye)	Dimroth and Reichardt (1969)
E_T^N	Normalized $E_T(30)$	Reichardt and Harbusch-Görnert (1983)
E_T^{SO}	Polarity ($n \rightarrow \pi^*$ abs freq of a sulfoxide)	Walter and Bauer (1977)
G	Polarity (IR shifts)	Allerhand and Schleyer (1963)
P	Polarity (^{19}F nmr shifts)	Brownlee *et al.* (1972)
P_y	Polarity ($\pi^* \rightarrow \pi$ emission of pyrene: relative intensities of two bands)	Dong and Winnick (1984)
S	Polarity (composite)	Brownstein (1960)
S'	Polarity (composite)	Drago (1992)
SPP	Polarity (Difference between UV-VIS spectra of 2-(dimethylamino)-7-nitrofluorene and 2-fluoro-7-nitrofluorene)	Catalán *et al.* (2001)
Z	Polarity (CT trans. of a pyridinium iodide)	Kosower (1958)
δ	Polarizability correction to π^*	Kamlet *et al.* (1983)
θ_{1k}, θ_{2k}	Polarity and polarizability components of π^*	Sjöström and Wold (1981)
π^*	Polarity/polarizability	Kamlet *et al.* (1977)
π^*_{azo}	Polarity ($n \rightarrow \pi^*$ and $\pi \rightarrow \pi$ shifts in six azo-merocyanines)	Buncel and Rajagopal (1989, 1990)
χ_R, χ_B	Polarity ($\pi \rightarrow \pi^*$ trans. of merocyanine dyes)	Brooker *et al.* (1965)
Φ	Polarity ($n \rightarrow \pi^*$ trans. in ketones)	Dubois and Bienvenüe (1968)
Part 2: Parameters based on equilibrium (physical or chemical)		
I	Gas-chromatographic retention index	Kováts (1965)
$K_{o/w}$ (as $\log_{10}$)	Lyophilicity (partition between 1-octanol and water)	Hansch (1969), Leo (1983)
KB	Kauri-butanol number (turbidity in solution of kauri resin in 1-butanol on addition of solvent)	ASTM (1982)
L	Desmotropic constant (enol/diketo ratio of ethyl acetoacetate in the solvent)	Meyer and Hopff (1921)
M	Miscibility number	Godfrey (1972)
$P_{o/w}$	(Same as $K_{o/w}$; *vide supra*.)	
$\mathscr{S}$ (script S)	Ionizing power ($\log k_2$ for a Menschutkin rn.)	Drougard and Decroocq (1969)
X	Polarity (an S_E2 reaction)	Gielen and Nasielski (1967)
Y	Ionizing power (t-butyl chloride solvolysis)	Grunwald and Winstein (1948)
Y_X	Ionizing power (various solvolyses)	Various authors
D_1	Conformational equilibrium shift	Eliel and Hofer (1973)
Ω	Polarity ([endo]/[exo] ratio in a Diels–Alder rn.)	Berson *et al.* (1962)

Table A.2a (*Contd*)

Symbol	Measure of (Basis)	Author, Ref.
Part 3: Parameters based on other properties		
$g(\epsilon_r)$	Polarity/polarizability ($=(\epsilon_r-1)/(2\epsilon_r+1)$; ϵ_r is the relative permittivity (dielectric constant).	Kirkwood (1934)
P	Polarizability $=(n^2-1)/(n^2+2)$	Koppel and Pal'm (1972) (Lorentz-Lorenz 1880)
Q	Polarity: $P-R$	Koppel and Pal'm (1972)
R	Equal to Y: see below	Koppel and Pal'm (1972)
SP	Solvophobic power (Gibbs energy of transfer of a nonpolar solute from water to the solvent)	Abraham *et al.* (1988)
Y	Polarizability/polarity: $=(\epsilon_r-1)/(\epsilon_r+2)$ (cf. Kirkwood)	Koppel and Pal'm (1972) (Mossotti 1850, Clausius 1879)
δ_H	Solubility parameter (square root of cohesive energy density $\Delta E_{vap}/V_m$)	Hildebrand and Scott (1950)
(No symbol)	Softness (used with Lewis acid/base strengths)	Pearson (1963)
$^1\chi^v/f$	1st-order valence molecular connectivity index	Kier (1981)
S_{orb}	Softness: $=1/(E_{LUMO}-E_{HOMO})$	Klopman (1968)

Table A.2b Dual Parameters

Acid	Measure of (basis)	Basic	Measure of (basis)	Author, ref.
E	Electrophilic solvation. ($E_T(30)$ corrected for non-specific effects: polarity and polarizability)	B	Nucleophilic solvation. (O–D IR stretch frequency difference, solvent-gas phase)	Koppel and Pal'm (1972); Koppel and Paju (1974)
A_j	Acity: anion-solvating tendency (see B_j)	B_j	Basity: cation-solvating tendency. (Bilinear correlation (with A_j) of many effects.)	Swain *et al.* (1983)
AN	Electron pair acceptor number (^{31}P NMR chemical shifts in triethylphosphine oxide)	DN	Pair donor number ($-\Delta H$ of adduct formation with SbF_5 in dilute solution in 1,2-dichloroethane)	Mayer *et al.* (1975, 1977; Gutmann (1976))
α	HBD strength (enhanced solvatochromism of Reichardt's dye 30 relative to 4-nitroanisole)	β	HBA strength (enhanced solvatochromism in HBA solvents for 4-nitroaniline relative to *N,N*-diethy1-4-nitroaniline)	Kamlet and Taft (1976)
C_A	Covalent part of acid strength (fit of enthalpy of adduct formation between EPD and EPA in solution to $-\Delta H = C_A C_B + E_A E_B$)	C_B	Covalent part of base strength (see C_A)	Drago and Wayland (1965); Drago (1980)
E_A	Electrovalent part of acid strength (see C_A)	E_B	Electrovalent part of base strength (see C_A)	"
		N	Nucleophilicity (bilinear correlation (with *Y, vide supra*) of rates of solvolysis of *t*-butyl chloride)	Winstein *et al.* (1954)
		D_H	Hard basicity (Gibbs energy of transfer of Na^+, water to solvent)	Persson *et al.* (1987)
		D_S	Soft basicity (wavenumber shift of sym. stretch of $HgBr_2$)	Persson *et al.* (1987)
		D_π	Soft basicity (retardation of [2+2] cycloaddiotion of diazodiphenylmethane to tetracyanoethene)	Oshima and Nagai (1985)
E_B^N	Lewis acidity (solvatochromism of $n \rightarrow \pi^*$ transition in 2,2,6,6,tetramethylpiperidine-1-oxide radical)			Mukerjee *et al.* (1982); Janowski *et al.* (1985)
		$\Delta H^0_{D\text{-}BF3}$	Hard basicity (enthalpy of adduct formation between EPD solvent and BF_3 in dilute solution in dichloromethane)	Maria and Gal (1985)
		μ	Base softness (difference in Gibbs energies of transfer of Ag^+ and alkali metal ion from water to solvent)	Marcus (1987)
α	Soft acidity (partial correlation of log K vs. E_n for Lewis acid/base reactions of cations with ligand Y^-)	E_n	Soft basicity ($=E^0+2.60$, where E^0 is the standard oxidation potential for $Y^- = \frac{1}{2}Y_2 + e^-$)	Edwards (1956)
β	Hard (protonic) acidity (partial correlation of log K, (as preceding) vs. H.	H	$1.74 + pK$ for $HY = H^+ + Y^-$	Edwards (1956)
SA	Comparison of solvatochromism of an unhindered with a sterically hindered stilbazolium betaine dye.	SB	Comparison of solvatochromism of 5-nitroindoline with *N*-methyl-5-nitroindoline	Catalán (2001)

Table A.3 Values of selected parameters for selected solvents

Solvents	1. R_v	2. Q_v	3. $\beta_\mu^{1/2}$	4. δ_H	5. E_T^N	6. Z	7. A_j	8. B_j	9. α	10. β	11. AN
1. n-hexane	0.229	0.00	0.00	14.9	0.009	50.0	0.00	0.00	0.00	0.00	0.0
2. Cyclohexane	0.256	0.00	0.00	16.9	0.006	60.1	0.02	0.06	0.00	0.00	—
3. Benzene	0.294	0.00	0.00	18.8	0.111	54.0	0.15	0.59	0.00	0.10	8.2
4. Toluene	0.293	0.03	0.50	18.2	0.099	54.0	0.13	0.54	0.00	0.11	—
5. Tetrachloromethane	0.274	0.03	0.00	17.6	0.052	52.0	0.09	0.34	0.00	0.00	8.6
6. 1,2-dichloroethane	0.266	0.50	3.56	20.3	0.327	63.4	0.30	0.82	—	—	16.7
7. Chlorobenzene	0.306	0.31	2.80	19.4	0.188	58.0	0.20	0.65	0.00	0.07	—
8. Acetonitrile	0.212	0.71	8.53	24.2	0.460	71.3	0.37	0.86	0.19	0.31	18.9
9. Diethyl ether	0.216	0.32	1.95	15.4	0.117	55.0	0.12	0.34	0.00	0.47	3.9
10. Tetrahydrofuran	0.246	0.44	3.37	19.0	0.207	58.8	—	—	0.00	0.55	8.0
11. 1,4-dioxane	0.254	0.04	0.85	20.5	0.164	64.6	0.19	0.67	0.00	0.37	10.8
12. Acetone	0.221	0.65	5.49	19.7	0.355	65.5	0.25	0.81	0.08	0.48	12.5
13. 2-butanone	0.231	0.63	5.08	19.0	0.327	64.0	0.23	0.74	—	—	—
14. Ethyl acetate	0.227	0.40	3.22	18.4	0.228	59.4	0.21	0.59	0.00	0.45	9.3
15. N,N-dimethylformamide	0.258	0.66	6.43	24.0	0.386	68.4	0.30	0.93	0.00	0.69	16.0
16. Hexamethylphosphoramide	0.273	0.63	7.31	18.3	0.315	62.8	0.00	1.07	0.00	1.05	10.6
17. Dimethyl sulphoxide	0.283	0.66	8.38	26.6	0.444	70.2	0.34	1.08	0.00	0.76	19.3
18. Pyridine	0.299	0.50	4.60	21.6	0.302	64.0	0.24	0.96	0.00	0.64	14.2
19. Triethylamine	0.243	0.09	1.28	15.2	0.043	50.0	0.08	0.19	0.00	0.71	1.43
20. Nitromethane	0.232	0.68	8.49	25.8	0.481	71.2	0.39	0.92	—	—	20.5
21. Dichloromethane	0.255	0.47	3.40	20.3	0.309	64.2	0.33	0.80	0.30	0.00	20.4
22. Chloroform	0.267	0.30	2.21	18.9	0.259	63.2	0.42	0.73	0.44	0.00	23.1
23. Methanol	0.204	0.71	4.68	—	0.762	83.6	0.75	0.50	0.93	0.62	41.3
24. Ethanol	0.221	0.66	3.97	—	0.654	79.6	0.66	0.45	0.83	0.77	37.9
25. 1-propanol	0.235	0.63	3.32	—	0.617	78.3	0.63	0.44	0.78	—	37.3
26. 2-propanol	0.230	0.63	3.29	—	0.546	76.3	0.59	0.44	0.76	0.95	33.6
27. 1,2-ethanediol	0.257	0.67	4.14	—	0.790	85.1	0.78	0.84	—	—	—
28. Water	0.206	0.76	7.27	—	1.000	94.6	1.00	1.00	1.17	0.47	54.8
29. Carbon disulphide	0.355	0.00	0.00	20.4	0.065	52.0	0.10	0.38	—	—	—

12. ***DN***	13. E_b	14. C_b	15. π^*	16. π^*_{azo}	17. **A**(^{14}N)	18. ***S***	19. χ_R	20. $\mathcal{S}$	21. ***W***	22. Ω	23. μ	24. log $K_{o/w}$
5.3	—	—	−0.08	−0.09	1.513	−0.337	50.9	—	—	—	—	4.11
4.8	—	—	0.00	0.00	—	−0.324	50.0	−4.15	—	0.595	—	3.44
—	0.143	1.360	0.59	0.40	1.540	−0.215	46.9	−1.74	—	0.497	—	2.13
—	0.087	1.910	0.54	0.38	1.535	−0.237	47.2	—	—	—	0.350	2.69
9.6	—	—	0.28	0.15	1.533	−0.245	48.7	−2.85	—	—	0.150	2.64
0.0	—	—	0.81	0.63	1.566	−0.151	—	−0.42	—	0.600	0.030	1.48
—	—	—	0.71	0.58	1.547	−0.182	45.2	−1.15	—	—	0.250	2.80
14.1	0.890	1.340	0.75	0.63	1.567	−0.104	45.7	−0.33	−4.221	0.692	0.350	−0.34
19.2	1.340	2.850	0.27	0.16	1.533	−0.277	48.3	−2.92	−7.300	0.466	0.050	0.89
20.0	0.978	4.270	0.58	0.40	1.537	—	46.6	−1.54	−6.073	—	—	—
14.8	1.090	2.380	0.55	0.34	1.545	−0.179	48.4	−1.43	—	—	0.050	−0.27
17.0	0.987	2.330	0.71	0.53	1.553	−0.175	45.7	−0.82	−5.067	0.619	0.030	−0.24
17.4	—	—	0.67	0.61	—	—	—	—	—	—	—	—
17.1	0.975	1.740	0.55	0.37	—	−0.210	47.2	−1.66	−5.947	—	—	0.73
26.6	1.230	2.480	0.88	0.86	1.564	−0.142	43.7	−0.22	−4.298	0.620	0.110	—
38.8	1.730	1.330	0.87	0.85	—	—	—	—	—	—	0.290	—
29.8	0.963	3.250	1.00	1.00	1.569	—	42.0	—	−3.738	—	0.220	−1.35
33.1	1.170	6.400	0.87	0.80	1.561	−0.197	43.9	—	−4.670	—	0.640	0.65
0.5	0.808	11.540	0.14	0.05	—	−0.285	49.3	—	—	0.445	—	1.64
2.7	—	—	0.85	0.70	1.576	−0.134	44.0	0.04	−3.921	0.680	0.030	—
—	—	—	0.82	0.62	1.575	−0.189	44.9	−0.55	—	—	—	1.25
—	—	—	0.58	0.62	1.586	−0.200	44.2	−0.89	—	—	0.150	1.97
18.4	0.780	1.120	0.60	0.60	1.621	0.050	43.1	−1.89	−2.796	0.845	0.020	−0.77
—	—	—	0.54	0.54	1.603	0.000	43.9	−2.02	−3.204	0.718	0.080	−0.31
—	—	—	0.52	0.52	—	−0.016	44.1	—	—	—	0.160	0.25
—	—	—	0.48	0.51	1.597	−0.041	44.5	—	−3.970	—	—	0.05
—	—	—	0.64	0.62	1.636	0.068	40.4	—	—	—	−0.030	−1.36
—	—	—	1.09	1.03	1.717	0.154	—	—	−1.180	0.869	0.000	—
—	—	—	0.30	0.25	1.529	−0.240	—	—	—	—	—	—

References

Abboud, J.-L. M., Notario, R., Bertrán, J., and Solà, M. (1993). *Prog. Phys. Org. Chem.*, **19**, 1–182.

Abboud, J.-L. M. and Taft, R. W. (1979). *J. Phys. Chem.*, **83**, 412–419.

Abboud, J.-L. M., Taft, R. W., and Kamlet, M. J. (1985). *J. Chem. Soc., Perkin Trans.*, **2**, 815–819

Abraham, M. H. (1982). *J. Am. Chem. Soc.*, **104**, 2085–2094.

Abraham, M. H. (1974). *Prog. Phys. Org. Chem.*, **11**, 1–87.

Abraham, M. H., Grellier, P. L., and McGill, R. A. (1988). *J. Chem. Soc., Perkin Trans.*, **2**, 339–345.

Adams, C., Earle, M., Roberts, G., and Seddon, K. (1998). *Chem. Commun.*, **19**, 2097–2098.

Ahlberg, P., Davidson, Ö., Löwendahl, M., Hilmersson, G., Karlsson, A., Håkansson, M. (1997). *J, Am. Chem. Soc.*, **119**, 1745–1750.

Ahrland, S., Chatt, J., and Davies, N. R. (1958). *Quart. Rev.* (London), **12**, 265–276.

Allerhand, A. and Schleyer, P. v. R. (1963). *J. Am. Chem. Soc.*, **85**, 371–380.

Alvarez, F. J. and Schowen, R. L. (1987). In Buncel, E. and Lee, C. C., eds. *Isotopes in organic chemistry*, Elsevier, Vol 7, pp 1–60.

American Society for Testing and Materials (1982). *Annual book of ASTM standards 1982*, Part 29, pp. 142–144 (ASTM D 1133–78), Philadelphia, PA.

Amis, E. S. and LaMer, V. K. (1939). *J. Am. Chem. Soc.*, **61**, 905–913.

Amis, E. S. and Hinton, J. F. (1973). *Solvent effects on chemical phenomena*, Vol. 1, Academic Press, New York and London.

Amis, E. S. and Price, J. E. (1943). *J. Phys. Chem.*, **47**, 338–348.

Arshadi, M., Yamdagni, R., and Kebarle, P. (1970). *J. Phys. Chem.*, **74**, 1475–1482.

Ashbaugh, H. S. and Paulaitis, M. E. (1996). *J. Phys. Chem.*, **100**, 1900–1913.

Atkins, P. W. (1998). *Physical chemistry*, 6th edn. Oxford University Press, Oxford, and Freeman, New York.

Audrieth, L. F. and Kleinberg, J. (1953). *Non-aqueous solvents*. Wiley, New York; Chapman and Hall, London.

Bagno, A. (1998). Winter School on Organic Chemistry, Bressanone, Italy.

Balakrishnan, V. K., Dust, J. M., vanLoon, G. W., and Buncel, E. (2001). *Can. J. Chem.*, **79**, 157–173.

Barrow, G. M. (1988). *Physical chemistry*, 5th edn. McGraw-Hill, New York, pp. 327–340.

Barthel, J. and Gores, H.-J. (1994). In Mamantov, G. and Popov, A. I., eds. *Chemistry of nonaqueous solutions: Current progress*, VCH, New York, Weinheim, Cambridge, pp. 6, 11.

Barton, A. F. M. (1975). *Chem. Rev.*, **75**, 731–753.

Basolo, F. and Pearson, R. G. (1967). *Mechanisms of inorganic reactions*, 2nd edn. Wiley, New York.

Bates, E. D., Mayton, R. D., Ntai, I., Davis, J. H. Jr. (2002). *J. Am. Chem. Soc.* **124**, 926–927.

Becke, A. D. (1997). *J. Chem. Phys.*, **107**, 8554–8560.

Becke, A. D. (2000). *J. Chem. Phys.*, **112**, 4020–4026.

Bekárek, V. (1983). *J. Chem. Soc., Perkin Trans.*, **2**, 1293–1296.

Bell, R. P., Bascombe, K. N. and McCoubrey, J. C. (1956). *J. Chem. Soc.*, 1286–1291.

Bell, R. P. (1941). *Acid-base catalysis*, Oxford University Press, London.

Bell, R. P. (1973). *The proton in chemistry*, 2nd edn. Cornell University Press, Ithaca, New York.

Bentley, T. W. and Roberts, K. (1985). *J. Org. Chem.*, **50**, 4821–4828.

Bernasconi, C.F. (1992). *Adv. Phys. Org. Chem.*, 27, 119–238.

Berry, R. S., Rice, S. A., and Ross, J. (2000). *Physical chemistry*. Oxford University Press, New York and Oxford.

Berson, J. A., Hamlet, Z., and Mueller, W. A. (1962). *J. Am. Chem. Soc.*, **84**, 297–304.

Billmeyer, F. W., Jr. (1971). *Textbook of polymer science*, 2nd edn. Wiley-Interscience, New York, pp. 289, 358–364.

Bjerrum, N. (1926). *Kgl.Danske Vidensk.Selskab.*, **7**, No. 9. See also Davies (1962), *or* Harned and Owen (1950).

Blagoeva, I. B., Toteva, M. M., Ouarti, N., and Ruasse, M.-F. (2001). *J. Org. Chem.*, **66**(6), 2123–2130.

Blandamer, M. J. (1977). In Gold, V. and Bethell, D., eds. *Adv. Phys. Org. Chem.*, **14**, Acad. Press, London. 203–343.

Blandamer, M. J. and Burgess, J. (1982). *Pure Appl. Chem.*, **54**(12), 2285–2296.

Blandamer, M. J., Scott, J. M. W., and Robertson, R. E. (1985). *Prog. Phys. Org. Chem.*, **15**, 149–196.

Bloom, H. and Hastie, J. W. (1965). In Waddington, T.C., ed. *Non-aqueous solvent systems*. Academic Press, London, New York, p. 353ff.

Bohme, D. K. (1984). In Almosta Ferreira, M.A., ed. *Ionic processes in the gas phase*, Reidel, Dordrecht.

Bohme, D. K. and Mackay, G. I. (1981). *J. Am. Chem. Soc.*, **103**, 978–979.

Boon, J., Levitsky, J., Pflug, J., and Wilkes, J. (1986). *J. Org. Chem.*, **51**, 480–483.

Bordwell, F. G. (1988). *Acc. Chem. Res.*, **21**, 456–463.

Born, M. (1920). *Z. Phys.*, **1**, 45–48.

Born, M. and Green, H. S. (1949). *General Kinetic Theory of Liquids*. Cambridge University Press, Cambridge.

Bosnich B. (1967). *J. Am. Chem. Soc.*, **89**, 6143–6148.

Bowden, K. and Cook, R. S. (1968). *J. Chem. Soc. (B)*, 1529–1533.

Bowman, N. S., McCloud, G. T., Schweitzer, G. K. (1968) *J. Amer. Chem. Soc.*, **90**, 3848–3852.

Brooker, L. G. S., Craig, A. C., Heseltine, D. W., Jenkins, P. W., and Lincoln, L. L. (1965). *J. Am. Chem. Soc.*, **87**, 2443–2450.

Brownlee, R. T. C., Dayel, S. K., Lyle, J. L., Taft, R. W. (1972). *J. Am. Chem. Soc.* **94**, 7208–7209.

Brownstein, S. (1960). *Can. J. Chem.*, **38**, 1590–1596.

Buncel, E., and Lawton, B. T. (1965). *Can. J. Chem.,* **43**, 862–875.

Buncel, E., Symons, E. A., and Zabel, A. W. (1965). *Chem. Commun.*, 173–174.

Buncel, E. (1975*a*). *Acc. Chem. Res.*, **8**, 132–139.

Buncel, E. (1975*b*). *Carbanions. Mechanistic and isotopic aspects*, Elsevier, Amsterdam, p. 16.

Buncel, E. (2000). *Can. J. Chem.*, **78**, 1251–1271.

Buncel, E., Boone, C., and Joly, H. A. (1986). *Inorg. Chim. Acta*, **125**, 167–172.

Buncel, E., Chuaqui, C., and Wilson, H. (1980). *J. Org. Chem.*, **45**, 3621–3626.

Buncel, E., Chuaqui, C., and Wilson, H. (1982). *Int. J. Chem. Kinetics*, **14**, 823–837.

Buncel, E., Davey, J. P., Jones, J. R., and Perring, K. D. (1990). *J. Chem. Soc., Perkin Trans.*, **2**, 169–173.

Buncel, E. and Dust, J. M. (2003). *Carbanion Chemistry: structures and mechanisms*. American Chemical Society, Washington, DC, Oxford University Press.

Buncel, E., Dust, J. M., and Terrier, F. (1995). *Chem. Rev.*, **95**, 2261–2280.

Buncel, E. and Menon, B. C. (1979). *J. Org. Chem.*, **44**, 317–320.
Buncel, E., Menon, B. C., and Colpa, J. P. (1979). *Can. J. Chem.*, **57**, 999–1005.
Buncel, E. and Millington, J. P. (1965). *Can. J. Chem.*, **43**, 556–564.
Buncel, E., Cannes, C., Chatrousse, A.P., and Terrier, F. (2002a). *J. Am. Chem. Soc.*, *124*, 8766–8767.
Buncel, E., Onyido, I. (2002b). *J. Labelle. Cpd. Radiopharm.* **45**, 291–306.
Buncel, E. and Rajagopal, S. (1989) *J. Org. Chem.*, **54**, 798–809.
Buncel, E. and Rajagopal, S. (1990) *Acc. Chem. Res.*, **23**, 226–231.
Buncel, E. and Rajagopal, S. (1991) *Dyes and Pigments*, **17**, 303.
Buncel, E. and Symons, E. A. (1976). *J. Am. Chem. Soc.*, **98**, 656–660.
Buncel, E. and Symons, E. A. (1981). In Bertini, I., Lunazzi, L., and Dei, A., eds. *Advances in solution chemistry*. Plenum Press, New York, pp. 355–371.
Buncel, E. and Symons, E. A. (1986). In Zuckerman, J.J., ed. *Inorganic reactions and methods*, Vol. 1. VCH, Weinheim, pp. 69–75.
Buncel, E., Symons, E. A., Dolman, D., and Stewart, R. (1970). *Can. J. Chem.*, **48**, 3354–3357.
Buncel, E. and Wilson, H. (1977a). *Adv. Phys. Org. Chem.*, **14**, 133–202.
Buncel, E. and Menon, B. (1977b). *J. Am. Chem. Soc.*, **99**, 4457–4461.
Buncel, E. and Wilson, H. (1979). *Acc. Chem. Res.*, **12**, 42–48.
Buncel, E., Wilson, H. (1980). *J. Chem. Educ.*, **57**, 629–633.
Buncel, E., Wilson, H., and Chuaqui, C. (1982). *J. Am. Chem. Soc.*, **104**, 4896–4900.
Buncel, E., Lee, C. C., eds. (1987). *Isotopes in Organic Chemistry*, Vol. 1. Isotopes in Molecular Rearrangements.
Buncel, E., Saunders, W. H., eds. (1992). *Isotopes in Organic Chemistry*, vol. 8, 'Heavy Atom Isotope Effects', Elsevier, Amsterdam.
Buncel, E., Park, K.-T., Dust, J.M., Manderville, R.A. (2003). *J. Am. Chem. Soc.* **125**, 5388–5392.
Bunnett, J. F. and Olsen, F. P. (1966). *Can. J. Chem.*, **44**, 1899–1916, 1917–1931.
Burgess, J. and Pelizzetti, E. (1992). *Prog. React. Kinet.*, **17**, 1–170.
Burrell, H. (1955). *Interchem. Rev.*, **14**, 3–16, 31–46.
Caldin, E. F. and Hasinoff, B. B. (1975). *J. Chem. Soc. Faraday Trans. I*, **71**, 515–527.
Caldin, E. F. and Grant, M. W. (1973). *J. Chem. Soc., Faraday Trans. I*, **69**, 1648–1654.
Cann, N. M. and Patey, G. N. (1997). *J. Chem. Phys.*, **106**, 8165–8195.
Car, R. and Parrinello, M. (1985). *Phys. Rev. Lett.*, **55**, 2471–2474.
Carey, F. A. (1996). *Organic chemistry*, 3rd edn. McGraw Hill, New York.
Carroll, F. A. (1998). *Perspectives on structure and mechanism in organic chemistry*, Brooks/Cole Publishers, Pacific Grove.
Carlson, R., Lundstedt, T., and Albano, C. (1985). *Acta Chem. Scand. B*, **39**, 79–91.
Carmichael, H. (2000). *Chem. Britain*, Jan., 36–38.
Catalán, J. (2001). In Wypych, G., ed. *Handbook of solvents*. ChemTec, Toronto, New York, pp. 583–616.
Cerfontain, H. (1968). *Mechanistic aspects in aromatic sulfonation and desulfonation*. Wiley, New York.
Chandler, D. and Andersen, H. C. (1972). *J. Chem. Phys.*, **57**, 1930–1937.
Chandrasekhar, J. and Jorgensen, W. L. (1985). *J. Am. Chem. Soc.*, **107**, 2974–2975.
Chapman, N. B. and Shorter, J., eds. (1972, 1978) *Advances in linear free energy relationships; correlation analysis in chemistry—recent advances.*, Plenum, London.
Chau, P. L., Forester, T. R., and Smith, W. (1996). *Mol. Phys.*, **89**, 1033–1055.
Chauvin, Y., Mussman, L., and Olivier, H. (1995). *Angew. Chem. Int. Ed. Engl.*, **34**, 2698–2700.

Chen, T., Hefter, G., and Marcus, Y. (2000). *J. Solution Chem.*, **29**, 201–216.
Chum, H. L. and Osteryoung, R. A. (1981). In Inman D. and Lovering, D.G., eds. *Ionic Liquids*. Plenum, New York & London, pp. 407–423.
Clark, R. J. and Brimm, E. O. (1965). *Inorg. Chem.*, **4**, 651–654.
Claydon, J., Greeves, N., Warren, S., and Wothers, P. (2001). *Organic chemistry*. Oxford University Press, Oxford.
Coetzee, J. F. and Ritchie, C. D., eds. (1969, 1976). *Solute–solvent interactions*. Marcel Dekker, New York, London.
Coetzee, J. F. and Deshmukh, B. K. (1990) *Chem. Rev.*, **90**, 827–835.
Collard, D. M., Jones, A. G., and Kriegel, R. M. (2001). *J. Chem. Educ.*, **78**, 70–72.
Collins, C. J. and Bowman, N. S., eds. (1970). *Isotope Effects in Chemical Reactions*, Van Nostrand Reinhold, New York.
Covington, A. K. and Newman, K. E., 1976. In Furter, W.F. 'Thermodynamic behaviour of Electrolytes in Mixed Solvents', *Adv. Chem. Ser.*, **155**, *Am. Chem. Soc.*, Washington, 153–196.
Cox, B. G. and Webster, K. C. (1936). *J. Chem. Soc.*, 1635–1637.
Cox, B. G. (1994). *Modern liquid phase kinetics*. Oxford University Press, Oxford.
Cox, B. G. (1973). *Ann. Rep. Chem. Soc. (A)*, **70**, 249–274.
Cox, R. A. and Yates, K. (1981). *Can. J. Chem.*, **59**, 2116–2124.
Cox, R. A. and Yates, K. (1983). *Can. J. Chem.*, **61**, 2225–2243.
Cox, R. A. (1987). *Acc. Chem. Res.*, **20**, 27–31.
Cox, R. A. (2000). *Adv. Phys. Org. Chem.*, **35**, 1–66.
Cox, R. A., Krull, U. J., Thompson, M., and Yates, K. (1979). *Anal. Chim. Acta*, **106**, 51–57.
Cox, R. A. and Yates, K. (1978). *J. Am. Chem. Soc.*, **100**, 3861–3867.
Coxeter, H. S. M. (1961). *Introduction to geometry*. Wiley, New York and London, p. 128, Eqn. 8. 84.
Cram, D. J., Rickborn, B., and Knox, G. R. (1960). *J. Am. Chem. Soc.*, **82**, 6412–6113.
Cram, D. J. (1965). *Fundamentals of carbanion chemistry*. Academic Press, New York, Chapters 2–4.
Cram, D. J. and Cram, J. M. (1978). *Acc. Chem. Res.*, **11**, 8–14.
Davidson, E. R. and Feller, D. (1986). *Chem. Rev.*, **86**, 681–686.
Davies, C. W. and James J. C. (1948). *Proc. Royal Soc.*, **195A**, 116–123.
Davies, C. W. (1962). *Ion association*. Butterworths, London.
de Rege, P. J. F., Gladyz, J. A., and Horváth, I. (1997). *Science*, **276**, 776–779.
Debye, P. J. W. and Hückel, E. (1923). *Z. Phys.*, **24**, 185–206.
Desrosiers, N. and Desnoyers, J. E. (1976). *Can. J. Chem.*, **54**, 3800–3808.
Dewar, M. J. S. and Storch, D. M. (1989) *J. Chem. Soc. Perkin Trans.*, **2**, 877–885.
Dewar, M. J. S. and Storch, D. M. (1985) *J. Chem. Soc. Chem. Commun.*, 94–96.
Dewar, M. J. S. and Thiel, W. J. (1977). *J. Am. Chem. Soc.*, **99**, 4899–4906, 4907–4917.
Dewar, M. J. S. (1992). In Seeman, J.I., ed. *A semiempirical life*. Am. Chem. Soc., Washington.
Dewar, M. J. S., Zoebisch, E. G., Healy, E. F., and Stewart, J. J. P. (1985). *J. Am. Chem. Soc.*, **107**, 3902–3909.
Dimroth, K., Reichardt, C., Siepmann, T., and Bohlmann, F. (1963). *Justus Liebigs Ann. Chem.*, **661**, 1–37.
Dimroth, K. and Reichardt C. (1969). *Justus Liebigs Ann. Chem.*, **727**, 93–105.
Dong, D. C. and Winnik, M. A. (1984) *Can. J. Chem.*, **62**, 2560–2565.
Dove, M. F. A. and Clifford, A. F. (1971). In Jander, G., Spandau, H., Addison, C.C., eds. *Chemie in nichtwässrigen Lösungsmitteln*. Vol. I, Part 1. pp. 119–300 (in Engl.). Vieweg, Braunschweig; Pergamon, Oxford.

Doye, J. P. K. and Wales, D. J. (1999). *Phys. Rev. B*, **59**, 2292–2300.

Drago, R. S., Ferris, D. C., and Wong, N. (1990). *J. Am. Chem. Soc.*, **112**, 8953–8961.

Drago, R. S. (1980). *Pure Appl. Chem.*, **52**, 2261–2274.

Drago, R. S. (1992). *J. Chem. Soc. Perkin Trans.*, **2**, 1827–1837.

Drago, R. S. and Kabler, R. A. (1972). *Inorg. Chem.*, **11**, 3144–3145.

Drago, R. S., Wong, N., and Ferris, D. C. (1991). *J. Am. Chem. Soc.*, **113**, 1970–1977.

Drago, R. S. and Wayland, B. B. (1965). *J. Am. Chem. Soc.*, **87**, 3571–3577.

Drljaca, A., Hubbard, C, D., van Eldrik, R., Asano, T., Basilevsky, M. V., and le Noble, W. J. (1998). *Chem. Rev.*, **98**, 2167–2289.

Drougard, Y. and Decroocq, D. (1969). *Bull. Soc. Chim. Fr.*, 2972–2983.

Dubois, J.-E. and Bienvenue, A. (1968) *J. Chim. phys.*, **65**, 1259–1265.

Dunn, W. J. III., Wold, S., Edlund, U., Hellberg, S., and Gasteiger, J. (1984). *Quantitative structure-activity relationships*, **3**, 1313–1317.

Dunn, E. J. and Buncel, E. (1989). *Can. J. Chem.*, **67**, 1440–1448.

Durrell, S. R. and Wallqvist, A. (1996). *Biophys. J.*, **71**, 1696–1706.

Dutkiewicz, M. (1990). *J. Chem. Soc., Faraday Trans.*, **86**, 2237–2241.

Dye, J. L. (1964). In *Solutions métal-ammoniac: propriétés physicochimiques*. pp. 137–145. Univ. Cath. Lille; Benjamin, New York.

Dyson, P. J., Ellis, D. J., Parker, D. G., and Welton, M. T. (1999). *Chem. Commun.*, (1), 25–26.

Džidić, I. and Kebarle, P. (1970). *J. Phys. Chem.*, **74**, 1466–1474.

Earle, M. J., McCormac, P. B. and Seddon, K. R. (1998), *Chem. Commun.*, 2245–2246.

Edwards, J. O. (1954). *J. Am. Chem. Soc.*, **76**, 1540–1547.

Edwards, J. O. (1956) *J. Am. Chem. Soc.*, **78**, 1819–1820.

Eigen, M. and Tamm, K. (1962). *Z. Electrochem.*, **66**, 107–121.

Eisfeld, W. and Regitz, M. (1996). *J. Am. Chem. Soc.*, **118**, 11918–11926.

El Hosary, A. A., Kerridge, D. H., and Shams El Din, A. M. (1981) In Inman D. and Lovering, D.G., eds. *Ionic liquids*. Plenum, New York & London, pp. 339–362.

Eliasson, B., Johnels, D., Wold, S., Edlund, U., and Sjöström, M. (1982). *Acta Chim. Scand.*, B, **36**, 155–164.

Eliel, E. L. and Hofer, O. (1973). *J. Am. Chem. Soc.*, **95**, 8041–8045.

Eliel, E. E., Wilen, S. H., and Mander, L. N. (1994). *Stereochemistry of organic compounds*. Wiley-Interscience, New York, pp. 416–421.

Ellis, B., Keim, W., and Wasserscheid, P. (1999). *Chem. Commun.*, **4**, 337–338.

Erdey-Grúz, T. (1974). *Transport phenomena in aqueous solutions*, Halstead (Wiley), New York.

Étard, A. (1881). *Ann. Chim. Phys.*, 5, **22**, 218–286.

Evans, M. G. and Polanyi, M. (1935). *Trans. Faraday Soc.*, **31**, 875–894.

Ewald, P. (1921). *Ann. Phys.*, **64**, 253–287.

Eyring, H. (1935). *J. Chem. Phys.*, **3**, 107–115.

Fabre, P.-L., Devynck, J., and Trémillon, B. (1982). *Chem. Rev.*, **82**, 591–614.

Fainberg, A. S. and Winstein, S. (1956). *J. Am. Chem. Soc.*, **78**, 2770–2777.

Fajans, K. (1923). See (1965) *Chem. Eng. News*, **43**, 96.

Fang, Y. R., Lai, Z. G. and Westaway, K. C. (1998). *Can. J. Chem.*, **76**, 758.

Fannin, A. A., Jr., Floreani, D. A., King, L. A., Landers, J. S., Piersma, B. J., Stech, D. J., Vaughn, R. L., Wilkes, J. S., and Williams, J. L.. (1984). *J. Phys. Chem.*, **88**, 2614–2621.

Feng, D.-F. and Kevan, L. (1980). *Chem. Rev.*, **80**, 1–20.

Fernelius, W. C. and Bowman, G. E. (1940). *Chem. Rev.*, **26**, 3–48.

Flood, H., Förland, T., and Motzfeldt, K. (1952). *Acta Chim. Scand.*, **6**, 257–269.
Foresman, J. B. and Frisch, Æ. (1993). *Exploring chemistry with electronic structure methods: A guide to using Gaussian*. Gaussian, Inc., Pittsburgh.
Fowler, H. W., Fowler, F. G., and Sykes, J. B. (1976). *Concise oxford dictionary*, 6th edn. Oxford University Press, Oxford.
Freemantle, M. (1998). *Chem. Eng. News*. 30 March 30, 32–37; (2000). *Chem. Eng. News.*, 15 May, 37–50; (2001). *Chem. Eng. News.*, 1 Jan., 21–25.
Frémy, M. E. (1856). *Ann. Chim. Phys.*, **47**, 5–50.
Fuoss R. M. and Krauss, C. A. (1933). *J. Am. Chem. Soc.*, **55**, 1019–1028.
Gielen, M. and Nasielski, J. (1967). *J. Organomet. Chem.*, **7**, 273–280.
Gillespie, R. J., Peel, T. E., and Robinson, E. A. (1971) *J. Am. Chem. Soc.*, **93**, 5083–5087.
Gillespie, R. J. and Peel, T. E. (1973) *J. Am. Chem. Soc.*, **95**, 5173–5178.
Gillespie, R. J. and Robinson, E. A. (1965). In Waddington, T.C., ed. *Non-aqueous solvent systems*. Academic Press, London, p. 117.
Godfrey, M. B. (1972). *CHEMTECH*, 359–363.
Gold, V. (1976). *J. Chem. Soc., Perkin Trans.*, **2**, 1531–1532.
Grant, G. H. and Richards, W. G. (1995). *Computational chemistry*. Oxford University Press, Oxford, New York, Toronto.
Gronwall, T. H., LaMer, V. K., and Sandved, K. (1928). *Z. Phys.*, **29**, 358–393.
Grunwald, E. and Winstein, S. (1948). *J. Am. Chem. Soc.*, **70**, 841–854.
Guillot, B., Guissani, Y., and Bratos, S. (1991). *J. Chem. Phys.*, **95**, 3643–3648.
Guillot, B. and Guissani, Y. (1993). *J. Chem. Phys.*, **99**, 8075–8094.
Guthrie, J.P. (1998). *J. Am. Chem. Soc.*, **120**, 1688–1694
Guthrie, J.P. and Guo, J. (1996). *J. Am. Chem. Soc.*, **118**, 11472–11487
Guterman, L. (1999). *Chemistry* (Summer), pp. 12–14. Am. Chem. Soc., Washington.
Gutmann, V. (1967). *Coord. Chem. Rev.*, **2**, 239–256.
Gutmann, V. (1969) *Chimia*, **23**, 285–292.
Gutmann, V. (1968). *Coordination chemistry in non-aqueous solutions* Springer-Verlag, Vienna & New York.
Gutmann, V. (1976) *Electrochim. Acta* **21**, 661–670.
Gutmann, V. and Scherhaufer, A. (1968). *Monatsh. Chem.*, **99**, 335–339.
Gutmann, V. and Wychera, E. (1966). *Inorg. Nucl. Chem. Lett.*, **2**, 257–260.
Haberfield, P. (1971). *J. Am. Chem. Soc.*, **93**, 2091–2093.
Haberfield, P., Lux, M. S., and Rosen, D. (1977). *J. Am. Chem. Soc.*, **99**, 6828–6831.
Hall, N. F. and Conant, J. B. (1927). *J. Am. Chem. Soc.*, **49**, 3047–3061; 3062–3070.
Hammett, L. P. (1970). *Physical organic chemistry*, 2nd edn. McGraw-Hill, New York.
Hammett, L. P. and Deyrup, (1932). *J. Am. Chem. Soc.*, **54**, 2721–2739.
Hammett, L. P. (1937). *J. Am. Chem. Soc.*, **59**, 96–103.
Hansch, C. (1969). *Acc. Chem. Res.*, **2**, 232–239.
Hansen, J. P. and McDonald, I. R. (1986). *Theory of simple liquids*. Academic Press (Harcourt Brace Jovanovich) London.
Hantzsch, A. and Caldwell, K. S. (1908). *Z. Phys. Chemie*, **61**, 227–240.
Hao, C. and March, R. E. (2001). *J. Mass Spect.*, **36**, 79–96.
Hao, C., March, R. E., Croley, T. R., Smith, J. C., and Rafferty, S. P. (2001). *J. Mass Spect.*, **36**, 79–96.
Harris, J. M., Shafer, S. G., Moffatt, J. R. and Becker, A. R. (1979). *J. Am. Chem. Soc.*, **101**, 3295–3300.
Hehre, W. J., Radom, L., Schleyer, P. v. R., and Pople, J. A. (1986). *Ab initio molecular orbital theory*. Wiley, New York.

Heirl, P. M., Ahrens, A. F., Henchman, M. J., Viggiano, A. A., and Paulson, J. F. (1988). *Faraday Discuss. Chem. Soc.*, **85**, 37–51.

Hildebrand, J. H. and Scott, R. L. (1950). *The solubility of nonelectrolytes*, 3rd edn. Reinhold, New York.

Hildebrand, J. H. and Carter, J. M. (1930). *Proc. Nat. Acad. Sci.*, **16**, 285–288.

Hildebrand, J. H. and Scott, R. L. (1962). *Regular solutions*, Prentice-Hall, Englewood Cliffs.

Hinton, J. F. and Amis, E. S. (1971). *Chem. Rev.*, **71**, 627–674.

Hogen-Esch, T. E. and Smid, J. (1966). *J. Am. Chem. Soc.*, **88**, 307–318; 318–324.

Hohenberg, P. and Kohn, W. (1964). *Phys. Rev.*, **136B**, 864–871.

House, H. O. (1972). *Modern synthetic reactions*, Benjamin, Menlo Park, CA.

Howarth, J., Hanlon, K., Fayne, D., and McCormac, P. (1997). *Tetrahedron Lett.*, **38**, 3097–3100.

Hughes, E. D. and Ingold, C. K. (1935). *J. Chem. Soc.*, 244–255.

Huheey, J. A., Keiter, E. A., and Keiter, R. L. (1993). *Inorganic chemistry*, 4th edn. HarperCollins College Publishers, New York.

Hunt, J. P. (1963). *Metal ions in solution*. Benjamin, New York, pp. 14–17, 27–35.

Hurley, F. H. and Weir, J. P. (1951). *J. Electrochem. Soc.*, **98**, 203–206.

Ingold, C. K. (1969). *Structure and mechanism in organic chemistry*, 2nd edn. Cornell University Press, Ithaca, New York.

Isaacs, N. S. (1984). In Buncel, E. and Lee, C.C., eds. *Isotopes in organic chemistry*, Vol. 6, Elsevier, Amsterdam.

Jander, J. (1966). *Anorganische und allgemeine Chemie in flüssigem Ammoniak*. Vol. I, part I of Jander, G., Spandau, H. and Addison, C.C., eds. *Chemie in nichtwässrigen ionisierenden Lösungsmitteln*. Vieweg, Braunschweig; Wiley/Interscience, New York and London.

Jander, J. and Lafrenz, Ch. (1970). *Ionizing solvents*. Verlag Chemie, Weinheim.

Janowski, A., Turowska-Tyrk, I., and Wrona, P. K. (1985). *J. Chem. Soc., Perkin Trans.*, **2**, 821–825.

Jencks, W.P. In 'Nucleophilicity', Harris, J.M. and McManus, S.P., eds. (1987). *ACS Advances in Chemistry Series*, American Chemical Society, Washington D.C., pp. 155–167 and references therein.

Jencks, D.A. and Jencks, W.P. (1977). *J. Am. Chem. Soc.*, **99**, 7848.

Jencks, W.P. (1969). *Catalysis in Chemistry and Enzymology*, McGraw-Hill, New York.

Johnson, D. E. (1998). *Applied multivariate methods for data analysts*. Brooks/Cole, Pacific Grove, Calif.

Jolly, W. L. (1970). *The synthesis and characterization of inorganic compounds*. Prentice-Hall, Englewood Cliffs.

Jones, J.R. (1973). *Prog. Reaction Kinetics*, *7*, 1.

Jorgensen, C. K. (1964). *Inorg. Chem.*, **3**, 1201–1202.

Jorgensen, W. L. (1982). *Chem. Phys. Lett.*, **92**, 405–410.

Jungers, J. C., Sajus, L., de Aguirre, I., and Decroocq, D. (1968). *L'Analyse cinétique de la transformation chimique*. Technip, Paris.

Kamlet, M. J., Abboud, J.-L. M., Abraham, M. H., and Taft, R. W. (1983). *J. Org. Chem.*, **48**, 2877–2887.

Kamlet, M. J., Minesinger, R. R., and Gilligan, W. H. (1972). *J. Am. Chem. Soc.*, **94**, 4744–4746.

Kamlet, M. J. and Taft, R. W. (1979) *J. Chem. Soc., Perkin Trans.*, **2**, 337–341, 349–356.

Kamlet, M. J. and Taft, R. W. (1976). *J. Am. Chem. Soc.*, **98**, 377–383; 2886–2894.

Kamlet, M. J., Jones, M. E., and Taft, R. W. (1979). *J. Chem. Soc. Perkin Trans.*, **2**, 342–348.

Kamlet, M. J., Abboud, J.-L. M., and Taft, R. W. (1981). *Prog. Phys. Org. Chem.*, **13**, 485–630.

Kamlet, M. J., Abboud, J.-L. M., and Taft, R. W. (1977) *J. Am. Chem. Soc.*, **99**, 6027–6038.

Kamlet, M. J., Doherty, R. M., Abraham, M. H., Carr, P. W., Doherty R. F., and Taft, R. W. (1987). *J. Phys. Chem.*, **91**, 1996–2004.

Kebarle, P. (1972). In Szwarc, M., ed. *Ions and ion pairs in organic reactions*, Vol. 1, Wiley, New York, Chapter 2.

Kerridge, D. H. (1978). In Lagowski, J. J., ed. *The chemistry of nonaqueous solvents*, Vol. VB., Academic Press, New York, pp. 269–329.

Kessler, Y. M., Puhovski, Y. P., Kiselev, M. G., and Vaisman, I. I. (1994) In Mamantov, G. and Popov, A. I., eds. *Chemistry of nonaqueous solutions: Current progress*, pp. 307–373. VCH, New York, Weinheim, Cambridge.

Kettle, S. F. A. (1996). *Physical inorganic chemistry*. pp. 331–335. Spektrum, Oxford.

Kier, L. B. (1981). *J. Pharm. Sci.*, **70**, 930–933.

Kilpatrick, M. and Jones, J. G. (1967). In Lagowski, J. J., ed. *The chemistry of nonaqueous solvents*, Vol. II., pp. 43–99. Academic Press, New York.

Kirkwood, J. G. (1934). *J. Chem. Phys.*, **2**, 351–361.

Kleinberg, R. and Brewer, P. (2001). *American Scientist*, **89**, 244–251.

Klemperer, W. (2001). *Science*, **293**, 815–816.

Klopman, G. (1968). *J. Am. Chem. Soc.*, **90**, 223–234.

Knauer, B. R. and Napier, J. J. (1974). *J. Am. Chem. Soc.*, **98**, 4395–4400.

Koch, W. and Holthausen, M. C. (2001). A chemist's guide to density functional theory. FVA-Frankfurter Verlag GmbH; Wiley, New York.

Koga, K., Tanaka, H., and Zeng, X. C. (1996). *J. Phys. Chem.*, **100**, 16711–16719.

Kohn, W. and Sham, L. J. (1965). *Phys. Rev.*, **140** A, 1133–1138.

Koo, I. S., An, S. K., Yang, K., Koh, H. J., Choi, M. H. and Lee, I. (2001). *Bull. Korean Chem. Soc.*, **22**(8), 842–846.

Koppel, I. A. and Paju, A. (1974) *Organic reactivity* (USSR). **11**, 121.

Koppel, I. A. and Pal'm, V. A. (1972). In Chapman, N.B. and Shorter, J., eds. *Advances in linear free energy relationships*. Plenum, New York & London, pp. 203ff.*

Kosower, E. M. (1958). *J. Am. Chem. Soc.*, **80**, 3253–3260.

Kosower, E. M. (1968). *An introduction to physical organic chemistry*. Wiley, New York.

Kováts, E. sz. (1965). *Adv. Chromatogr.*, **1**, 229–247

Kresge, A. J., More O'Ferrall, R. A. and Powell, M. F. (1987). In Buncel, E. and Lee, C. C., eds. *Isotopes in organic chemistry*, Elsevier, Vol. 7, pp 177–273.

Kumar, A. (2001). *Chem. Rev.*, **101**, 1–19.

Kusalik, P. G. and Svishchev, I. (1994). *Science*, **265**, 1219–1221.

Kyba, E. P., Koga, K., Sousa, L. R., Siegel, M. G., and Cram, D. J. (1973). *J. Am. Chem. Soc.*, **95**, 2692–2693.

Lagowski, J. J. and Moczygemba, G. A. (1967). In Lagowski, J. J., ed. *The chemistry of nonaqueous Solvents*, Vol. II., Academic Press, New York, pp. 319–371.

Lagowski, J. J. (1971). *Pure Appl. Chem.*, **25**, 429–464.

Laidler, K. J. and Meiser, J. M. (1995). *Physical chemistry*, 2nd edn. Houghton Mifflin, Boston.

Laidler, K. J. (1987). *Chemical kinetics*, 3rd edn. Harper and Row, New York.

Lander, J. and Lafrenz, Ch. (1970). *Ionizing solvents*, Wiley, Verlag Chemie, Weinheim, p. 139.

Langford, C. H. and Tong, J. P. K. (1977). *Acc. Chem. Res.*, **10**, 258–264.
Langhals, H. (1982*a*). *Nouv. Jour. Chim.*, **6**, 265–267.
Langhals, H. (1982*b*) *Angew. Chem. (Int Edn., Engl.)* B21, 724–733.
Latimer, W. M., Pitzer K. S., and Slansky, C. M. (1939). *J. Chem. Phys.*, **7**, 108–111.
Lau, Y. K., Ikuta, S., and Kebarle, P. (1982). *J. Am. Chem. Soc.*, **104**, 1462–1469; see also Kebarle, P. (1972), In Szwarc, M., ed. *Ions and ion Pairs in organic reactions*, Wiley-Interscience, New York, 51–83.
Leach, A. R. (1996). *Molecular modelling: principles and applications.* Addison Wesley Longman Limited, Harlow.
Leo, A. (1983) *J. Chem. Soc. Perkin Trans.*, **2**, 825–838.
Lewars, E. G. (2003). *Computational Chemistry.* Kluwer Academic Publishers, Boston, 388–389.
Levine, I. N. (2000). *Quantum chemistry*, 5th edn. Prentice Hall, Englewood Cliffs, NJ.
Lux, H. (1939). *Z. Elektrochem.*, **45**, 303–309.
Mackay, D., Shiu, W. Y., and Ma, K. C. (1992). *Illustrated handbook of physical-chemical properties and environmental fate of organic chemicals.* Lewis, Boca Raton.
Madan, B. and Sharp, K. (1996). *J. Phys. Chem.*, **100**, 7714–7721.
Maksimović, Z. B., Reichardt, C., and Spirić, A. Z. (1974). *Z. Anal. Chem.*, **270**, 100–104.
Malinowski, E. R. and Howery, D. G. (1980). *Factor analysis in chemistry.* Wiley, New York; reprinted 1989 by Krieger, Malabar (Fla.).
Manuta, D. M. and Lees, A. J. (1986). *Inorg. Chem.*, **25**, 3212–3218.
Marcus, Y. (1985). *Ion solvation.* Wiley, Chichester
Marcus, Y. (1998). *The properties of solvents.* Wiley, Chichester.
Marcus, Y. (1987). *J. phys. chem.*, **91**, 4422–4428.
Maria, P.-C. and Gal, J.-F. (1985). *J. phys. chem.*, **89**, 1296–1304.
Maria, P.-C., Gal, J.-F., de Franceschi, J., and Fargin, E. (1987). *J. Am. Chem. Soc.*, **109**, 483–492.
Marziano, N. C., Traverso, P. G., Tomasin, A., and Passerini, R. C. (1977). *J. Chem. Soc., Perkin Trans.*, **2**, 309–313.
Marziano, N. C., Cimino, G. M., and Passerini, R. C. (1973). *J. Chem. Soc., Perkin Trans.*, **2**, 1915–1922.
Mashima, M., McIver, R. R., Taft, R. W., Bordwell, F. G., and Olmstead, W. N. (1984). *J. Am. Chem. Soc.*, **106**, 2717–2718.
Mattson, A. (2002). *Science*, **298**, 759–760.
Matubayasi, N. and Levy, R. M. (1996). *J. Phys. Chem.*, **100**, 2681–2688.
Matyushov, D. V., Schmid, R., and Ladanyi, B. M. (1997). *J. Phys. Chem.*, **101**, 1035–1050.
Maurois, A. (1927). *Les discours du Dr. O'Grady*. Trans. Thurfrida Wake. (1965) edn. combined with *The Silence of Colonel Bramble*, The Bodley Head, London, p. 176.
Mayer, U., Gerger, W., and Gutmann, V. (1977). *Monatsh. Chem.*, **108**, 489–498.
Mayer, U., Gutmann, V., and Gerger, W. (1975). *Monatsh. Chem.*, **106**, 1235–1257.
Melander, L. and Saunders, W. H. (1980). *Reaction rates of isotopic molecules.* Wiley, New York.
Meng, E. C. and Kollman, P. A. (1996). *J. Phys. Chem.*, **100**, 11460–11470.
Metropolis, N., Rosenbluth, A. W., Rosenbluth, M. N., Teller, A. H., and Teller, E. (1953). *J. Chem. Phys.*, **21**, 1087–1092.
Meyer, K. H. and Hopff, H. (1921). *Ber. Deutsche Chem. Ges.*, **54**, 579–580.

Miertus, S., Scrocco, E., and Tomasi, J. (1981). *Chem. Phys.*, **55**, 117–129.

Mitchell, S. A. (1992). In Fontijn, ed. *Gas-phase metal reactions*, A. Elsevier, Amsterdam, pp. 227–252.

More O'Ferrall, R. A. (1970). *J. Chem. Soc. B.*, 274

More O'Ferrall, R. A., Koeppl, G. W. and Kresge, A. J. (1971). *J. Amer. Chem. Soc.*, **93**, 9–20.

Møller, C. and Plesset, M. S. (1934). *Phys. Rev.*, **46**. 618–622.

Moore, J. W. and Pearson, R. G. (1981). *Kinetics and mechanism*, 3rd edn. Wiley, New York.

Moore, W. J. (1972). *Physical chemistry*, 4th edn. Prentice-Hall, Englewood Cliffs, p. 571.

Morrison, R. T. and Boyd, R. N. (1992). *Organic Chemistry*, 6th edn. Preintice-Hall, Englewood Cliffs, NJ.

Mu, L., Drago, R. S. and Richardson, D. E. (1998). *J. Chem. Soc. Perkin Trans.*, **2**, 159–167.

Mukerjee, P., Ramachandran, C., and Pyter, R. A. (1982). *J. Phys. Chem.*, **86**, 3189–3197.

Nash, O. (1938). "Where There's a Will, There's Velleity". In *I'm a stranger here myself*, Little, Brown & Co., Boston.

Némethy, G. and Scheraga, H. A. (1962). *J. Chem. Phys.*, **36**, 3401–3417.

Newton, M. D. (1997). *J. Chem. Phys.*, **67**, 5535–5546.

Nigretto, J.-M. and Jozefowicz, M. (1978). In Lagowski, J. J., ed. *The chemistry of nonaqueous solvents*, vol. VA., Academic Press, New York, pp. 179–250.

Oh, Y. H., Jang, G. G., Lim, G. T. and Ryu, Z. H. (2002). *Bull. Korean Chem. Soc.*, **23**, 1089–1096.

Okazaki, S., Nakanishi, K., Touhara, H., and Adachi, Y. (1979). *J. Chem. Phys.*, **71**, 2421–2429.

Okazaki, S., Nakanishi, K., Touhara, H., Watanabi, N., and Adachi, Y. (1981). *J. Chem. Phys.*, **74**, 5863–5871.

Olah, G. A., Prakash, G. K. S., and Sommer, J. (1985). *Superacids*. Wiley, New York.

Onsager, L. (1936). *J. Am. Chem. Soc.*, **58**, 1486–1493.

Oshima, T. and Nagai, T. (1985). *Tetrahedron Lett.*, **26**, 4785–4788.

Parker, A. J. (1962). *Quart. Rev.* (London). **16**, 163–187.

Parker, A. J. (1967). *Adv. Phys. Org. Chem.*, **5**, 173–235.

Parker, A. J. (1969). *Chem. Rev.*, **69**, 1–32.

Passerini, R., Marziano, N. C., and Traverso, P. G. (1975). *Gazz. Chim. Ital.*, **105**, 901–906.

Peacock, S. C. and Cram, D. J. (1976). *J. Chem. Soc. Chem. Commun.*, 282–284.

Pearson, R. G., ed. (1973). *Hard and soft acids and bases*. Dowden, Hutchinson & Ross, Stroudsberg, PA.

Pearson, R. G. (1963). *J. Am. Chem. Soc.*, **85**, 3533–3539.

Pearson, R. G. (1988). *Inorg. Chem.*, **27**, 734–741.

Pellerite, M. J. and Brauman, J. I. (1980). In E. Buncel and T. Durst, eds., *Comprehensive carbanion chemistry*, vol. 5A, Elsevier, Amsterdam.

Percus, J. K. and Yevick, G. J. (1958). *Phys. Rev.*, **110**, 1–13.

Persson, I., Sandström, M., and Goggin, P. L. (1987). *Inorg. Chim. Acta*, **129**, 183–197.

Pethybridge, A. D. and Prue, J. E. (1972). *Prog. Inorg. Chem.*, **17**, 327–390.

Pine, S. H. (1987). *Organic chemistry*, 5th edn. McGraw-Hill, New York, etc.

Plešek, J. and Heřmánek, S. (1968). Trans. Mayer, K. *Sodium hydride: its use in the laboratory and in technology*. CRC Press, Cleveland.

Poos, G. I., Arth, G. E., Beyler, R. E., and Sarett, L. H. (1953). *J. Am. Chem. Soc.*, **75**, 422–429.

Pourbaix, M. J. N. (1949). *Thermodynamics of dilute aqueous solutions*. Arnold, London.
Pourbaix, M. J. N., Van Muylder, J., and de Zhoukov, N. (1963). *Atlas d'Equilibres électroniques à 25 °C*. Gauthier-Villars, Paris.
Pregel, M. J., Dunn, E. J., and Buncel, E. (1990). *Can. J. Chem.*, **68**, 1846–1858.
Pregel, M. J., Dunn, E. J., Negelkerke, R., Thatcher, G. R. J., and Buncel, E. (1995). *Chem. Soc. Rev.*, **24**, 449–455.
Pregel, M. J., Dunn, E. J., and Buncel, E. (1991). *J. Am. Chem. Soc.*, **113**, 3545–3550.
Rabinowitch, E. and Wood, W. C. (1936). *Trans. Faraday Soc.*, **32**, 1381–1387.
Ràfols, C., Rosés, M., and Bosch, E. (1997). *J. Chem. Soc., Perkin Trans.*, **2**, 234–248.
Rahimi, A. K. and Popov, A. I. (1976). *Inorg. Nucl. Chem. Lett.*, **12**, 703–707.
Rau, H. (1983). *Chem. Rev.*, **83**, 535–547.
Rauhut, G., Clark, T., and Steinke, T. (1993). *J. Am. Chem. Soc.*, **115**, 9174–9181.
Reichardt, C. (1994). *Chem. Rev.*, **94**, 2319–2358.
Reichardt, C. (1988). *Solvents and solvent effects in organic chemistry*, 2nd edn. VCH, Weinheim.
Reichardt, C. (2003). Solvents and solvent effects in organic chemistry, 3rd edn. Wiley-VCH, Weinheim.
Reichardt, C. (1971). *Justus Liebigs Ann. Chem.*, **752**, 64–67.
Reichardt, C. and Harbusch-Görnert, E. (1983). *Justus Liebigs Ann. Chem.*, **727**, 721–743.
Reichardt, C., Löbbecke, S., Mehranpour, A. M., and Schäfer, G. (1998). *Can. J. Chem.*, **76**, 686–694.
Richard, J. P. (1995). *Tetrahedron*, **51**, 1535
Ritchie, C. D. (1969). In Coetzee J.F. and Ritchie, C.D., eds. *Solute-solvent interactions*. Marcel Dekker, New York & London, Chapter 4.
Robinson, R. A. and Stokes, R. H. (1959). *Electrolyte solutions*. Butterworths, London.
Röllgen, F. W., Bramer-Wegner, E., and Buttering, L. (1984). *J. Phys. Colloq.*, **45** (Suppl. 12), C9.
Scatchard, G. (1931). *Chem. Rev.*, **8**, 321–333.
Scatchard, G. (1932). *Chem. Rev.*, **10**, 229–240.
Schön, I. (1984). *Chem. Rev.*, **84**, 287–297.
Scriven, E. F. V., Toomey, J. E., and Murugan, R. (1994). In Kirk-Othmer, *Encyclopedia of chemical technology*, 4th edn., v. 20, pp. 641–679.
Searles, S. K. and Kebarle, P. (1969). *Can. J, Chem.*, **47**, 2619–2627.
Shafirovitch, V., Dourandin, A. and Geacintov, N. E. (2001). *J. Phys. Chem. B*, **105**, 8431–8435.
Shaik, S. S., Schlegel, H. B., and Wolfe, S. (1992). *Theoretical aspects of physical organic chemistry: the S_N2 mechanism*. Wiley, New York.
Smith, M. B. and March, J. (2001). *March's Advanced Organic Chemistry*, Wiley, New York, etc.
Simkin, B. Ya., and Sheikhet, I. I. (1995). In T. J. Kemp, ed. *Quantum chemical and statistical theory of solutions: a computational approach*. Engl. Ed. Ellis Horwood, London.
Sjöström, M. and Wold, S. (1981). *J. Chem. Soc. Perkin Trans.*, **2**, 104–109.
Skoog, D. A., West D. M., and Holler, F. J. (1989). *Fundamentals of Analytical Chemistry*, Saunders, New York, pp. 124–130.
Smith, H. (1963). In *Organic reactions in liquid ammonia*. Vol. I, part 2 of Jander, G., Spandau, H., Addison, C.C., eds. *Chemie in nichtwäßrigen ionisierenden Lösungsmitteln*. Vieweg, Braunschweig; Wiley/Interscience, New York and London.

Smith, R. A. (1994). In Kirk-Othmer, *Encyclopedia of chemical technology*, 4th edn. Vol. 11, pp. 355–376.
Smithson, J. M. and Williams, R. J. P. (1958). *J. Chem. Soc.*, 457
Solomons, G. and Fryhle. C. (2000). *Organic chemistry*, 7th edn. Wiley, New York.
Song, C. E., Shim, W. H., Roh, E. J., and Choi, J. H. (2000). *Chem. Commun.*, **17**, 1695–1696.
Song, C. E., Shim, W. H., Roh, E. J., Lee, S., and Choi, J. H. (2001). *Chem. Commun.*, (12), 1122–1123.
Stairs, R. A. (1957). *J. Chem. Phys.*, **27**, 1431–1432.
Stairs, R. A. (1962). *Can. J. Chem.*, **40**, 1656–1659.
Stairs, R. A. (1976). In Furter, W. F., "Thermodynamic Behaviour of Electrolytes in Mixed Solvents", *Adv. in Chem. Ser.*, **155**, Am. Chem. Soc., Washington, 332–342.
Stairs, R. A. (1978) *Chem. Eng. News*, 5 June, p. 36 (letter).
Stairs, R. A. (1979). In Furter, W. F. "Thermodynamic Behaviour of Electrolytes in Mixed Solvents-II", *Adv. in Chem. Ser.*, **177**, Am. Chem. Soc., Washington., 167–176.
Stairs, R. A. (1983). *Chem 13 News* (Univ. of Waterloo, Canada). Oct., pp. 11–12.
Stevenson, C. D., Fico, R. M. Jr., and Brown, E. C. (1998). *J. Phys. Chem. B*, **102**, 2841–2844.
Stewart, R. (1985). *The proton: applications to organic chemistry*. Academic Press, Orlando.
Stillinger, F. H. and Rahman, A. (1974). *J. Chem. Phys.*, **60**, 1545–1557.
Strehlow, H. and Schneider, H., (1969). *J. Chim. Phys.*, **66**, 118–123.
Strehlow, H. and Schneider, H. (1971). *Pure Appl. Chem.*, **25**, 327–344.
Streitwieser, A., Heathcock, C. H., and Kosower, E. M. (1992). *Introduction to organic chemistry*. Macmillan, New York.
Suarez, P. A. Z., Dullius, J. E. L., Einloft, S., de Souza, R. F., and Dupont, J. (1997). *Inorg. Chim. Acta*, **255**, 207–209.
Svishchev, I., Kusalik, P. G., Wang, S., and Boyd, R. J. (1996). *J. Chem. Phys.*, **105**, 4742–4750.
Swain, C. G., Swain, M. S., Powell, A. L., and Alunni, S. (1983). *J. Am. Chem. Soc.*, **105**, 502–513.
Swieton, G., v. Jouanne, J., Kelm, H., and Huisgen, R. (1983). *J. Chem. Soc., Perkin Trans. II*, 37–43.
Symons, E. A. and Clermont, M. J. (1981). *J. Am. Chem. Soc.*, **103**, 3127–3130.
Symons, E. A., Clermont, M. J., and Coderre, L. A. (1981). *J. Am. Chem. Soc.*, **103**, 3131–3135.
Symons, E. A. and Buncel, E. (1972). *J. Am. Chem. Soc.*, **94**, 3641–3642.
Szwarc, M. (1968) *Carbanions, Living polymers and electron transfer processes*. Wiley-Interscience, New York, pp. 216–225.
Szwarc, M. (ed.) (1972). *Ions and ion pairs in organic reactions*. Wiley-Interscience, New York.
Taft, R. W., Abboud, J.-L. M., Kamlet, M. J., and Abraham, M. H. (1985). *J. Solution Chem.*, **14**, 153–175.
Taft, R. W. and Bordwell, F. G. (1988). *Acc. Chem. Res.*, **21**, 463–469.
Taft, R. W., Abboud, J.-L. M., and Kamlet, M. J. (1981). *J. Am. Chem. Soc.*, **103**, 1080–1086.
Tanaka, H. (1987). *J. Chem. Phys.*, **86**, 1512–1520.
Tapia, O. and Goscinski, O. (1975). *Mol. Phys.*, **29**, 1653–1661.
Terrier, F., McCormack, P., Kizilian, E., Halle, J.C., Demerseman, P., Guir, F., and Lion, C. (1991). *J. Chem. Soc. Perkin Trans.*, **2**, 153–158.

Terrier, F., Moutiers, G., Xiao, L., LeGuevel, E. and Guir, P. (1995). *J. Org. Chem.*, **60**, 1748–1754.

Thielmans, A. and Massart, D. l. (1985) *Chimia*, **39**, 236–242.

Thomson, B. A. and Iribarne, J. V. (1979). *J. Chem. Phys.*, **71**, 4451–4463.

Tobe, M. L. and Burgess, J. (1999). *Inorganic reaction mechanisms*. Addison Wesley Longman, Harlow.

Trémillon, B. (1971). *Pure Appl. Chem.*, **25**, 395–428.

Trémillon, B. (1974). *Chemistry in non-aqueous solvents*. Reidel, Dordrecht and Boston.

Troe, J. (1978). *Ann. Rev. Phys. Chem.*, **29**, 223–250.

Usanovich, M. (1939). *Zhur. Obshch. Khim.*, **9**, 182–192.

van Leeuwen, J. M. J., Groeneveld, J., and deBoer, J. (1959). *Physica*, **25**, 792–808.

van Eldik, R. and Hubbard, C. D. (eds.) (1997). *Chemistry under extreme or non-classical conditions*, Wiley, New York, Spektrum, Heidelberg, Chaps. 2–4.

van Eldik, R. and Meyerstein, D. (2000). *Acc. Chem. Res.*, **33**, 207–214.

van Eldik, R., Asano, T., and le Noble, W. J. (1989). *Chem. Rev.*, **89**, 549–688.

Vogel, A., revised. Bassett, J., Denney, R. C., Jeffrey, G. H., and Mendham, J. (1978). *Textbook of quantitative inorganic analysis*. Longman, Harlow (Essex), pp. 687–68.

Walden, P. (1914). *Bull. Acad. Imper. Sci.* (St. petersburg), 405–422.

Walter, W. and Bauer, O. H. (1977). *Justus Liebigs Ann. Chem.*, 421–429.

Walther, D. (1974). *J. Prakt. Chem.*, **316**, 604–614.

Wasserscheid, P., Gordon, C. M., Hilgers, C., Muldoon, M. J., and Dunkin, I. R. (2001). *Chem. Commun.*, **13**, 1186–1187.

Wasserscheid, P. and Keim, W. (2000). *Angew. Chem. Int. Ed. Engl.*, **39**, 3772–3789.

Welton, T. (1999). *Chem. Rev.*, **99**, 2071–2083.

Werblan, L., Rotowska, A., and Minc, S. (1971). *Electrochim. Acta*, **16**, 41–49.

Wheeler, C., West, K. N., Liotta, C. L., and Eckert, C. A. (2001). *Chem. Commun.*, **10**, 887–888.

Wilson, K. R. (1989). In Moreau, M. and Turq., P., eds. *Chemical reactivity in liquids: fundamental aspects*, Plenum, New York.

Winstein, S., Fainberg, A. H., and Grunwald, E. (1957), *J Am. Chem. Soc.*, **79**, 4146–4155.

Winstein, S., Clippinger, E., Fainberg, A. H., and Robinson, G. C. (1954) *J. Am. Chem. Soc.*, **76**, 2597–2598.

Wold, S. and Sjöström, M. (1978). In Shorter, J. and Chapman, N. B., eds. *Correlation analysis in chemistry: recent advances*. Plenum, New York, Chapter 1.

Wong, M. W., Wiberg, K. B., and Frisch, M. J. (1992). *J. Am. Chem. Soc.*, **114**, 1645–1642.

Wood, J. M., Hinchcliffe, P. S., Laws, A. P. and Page, M. I. (2002). *J. Chem. Soc., Perkin Trans.*, **2**, 938–946.

Wooley, E. M. and Hepler, L. G. (1972). *Anal. Chem.*, **44**, 1520–1523.

Wypych, G. (ed.) (2001). *Handbook of solvents*, ChemTec, Toronto and Wm. Andrew, Norwich, NY.

Yamane, H., Nakao, Y., Kawabe, S., Xie, Y., Kanehisa, N., Kai, Y., Kinoshita, M., Mon, L., and Hayashi, Y. (2001). *Bull. Chem. Soc. Japan*, **74**, 2107–2112.

Yarnell, J. L., Katz, M. J., Wenzel, R. G., and Koenig, S. H. (1973). *Phys. Rev. A*, **7**, 2130–2144.

Yingst, A. and McDaniel, D. H. (1967). *Inorg. Chem.*, **6**, 1067–1068.

Yu, H.-A., Pettitt, B. M., and Karplus, M. (1991). *J. Am. Chem. Soc.*, **113**, 2425–2434.

Index

QD
543
B945
2003
CHEM

RETURN TO: CHEMISTRY LIBRARY
100 Hildebrand Hall • 510-642-3753

LOAN PERIOD 1 1-MONTH USE	2	3
4	5	6

ALL BOOKS MAY BE RECALLED AFTER 7 DAYS.
Renewals may be requested by phone or, ~~using GLADIS, type inv followed by your patron ID number~~.

DUE AS STAMPED BELOW.

NON-CIRCULATING UNTIL: MAR 16 2004		
MAY 22 2004		
AUG 12 2005		
MAY 09 2006		
DEC 13		
AUG 16		

FORM NO. DD 10
2M 4-03

UNIVERSITY OF CALIFORNIA, BERKELEY
Berkeley, California 94720–6000